FRACKING PENNSYLVANIA

"*Fracking Pennsylvania* is an indispensable book for anyone who wants to understand natural gas fracking and the environmental risks it presents. But it is the politics of fracking, of how Gov. Tom Corbett essentially sold the state to the natural gas industry for beads and trinkets (and $1.8 million in donations to his political campaigns) that will cause the greatest outrage. Walter M. Brasch has done a public service by creating a devastating and coherent account of how Texas and Oklahoma energy kingpins got their way in Pennsylvania. The gas boom has brought lower natural gas and electricity prices, but at what cost to our future?"
—DAVID DeKOK, *Fire Underground: the Ongoing Tragedy of the Centralia Mine Fire*; *Unseen Danger: A Tragedy of People, Government, and Centralia*; *The Epidemic: A Collusion of Power, Privilege, and Public Health*.

"Walter Brasch has presented the many complicated issues surrounding shale gas drilling in Pennsylvania in a clear, easy to understand manner without oversimplifying them. Fracking Pennsylvania is packed with information every Pennsylvanian—and everyone living in any area being drilled or likely to be drilled by the gas industry—needs to know about the environmental, public health and safety, and economic risks this dangerous practice poses."
—Karen Feridun, founder, Berks Gas Truth

"Dr. Brasch has meticulously researched the issues behind hydraulic fracture mining (fracking) in all its complexity: geologic, environmental, social, economic and political. Shale gas being touted more and more as the greener energy of our future makes this book an essential reader for anyone who wants to safeguard their health and the environment."
—Carol Terracina Hartman, environmental journalist

"A remarkable piece of work, and a great resource for anyone working on the issue."
—Joe Uehlein, labor/environmental actisist and musician.

Fracking Pennsylvania:
Flirting With Disaster

by Walter M. Brasch

Greeley & Stone, Publishers, LLC
Carmichael, California 95608

LCCN 2012949557
ISBN 978-0-942991-16-1

Cover photo © 2010 International Wow Co./Docudrama Films

Design: MaryJayne Reibsome

PRINTED IN THE UNITED STATES OF AMERICA

Greeley & Stone, Publishers, LLC
4731 Whitney Ave., suite 20
Carmichael, Calif. 95608
www.greeleyandstone.com

Rig near Warrendale, Pa.

PHOTO: Gary F. Clark

BOOKS BY WALTER M. BRASCH

A Comprehensive Annotated Bibliography of
    American Black English (with Ila Wales Brasch)
Black English and the Mass Media
Cartoon Monickers: An Insight Into the Animation Industry
Columbia County Place Names
The Press and the State:
    Sociohistorical and Contemporary Interpretations
    (senior author)
A ZIM Self-Portrait
Forerunners of Revolution:
    Muckrakers and the American Social Conscience
With Just Cause: Unionization of the American Journalist
Enquiring Minds and Space Aliens:
    Wandering Through the Mass Media and Popular Culture
Social Foundations of the Mass Media (senior author)
The Joy of Sax: America During the Bill Clinton Era
Brer Rabbit, Uncle Remus, and the 'Cornfield Journalist':
    The Tale of Joel Chandler Harris
Sex and the Single Beer Can:
    Probing the Media and American Culture
America's Unpatriotic Acts:
    The Federal Government's Violation of
    Constitutional and Civil Rights
'Unacceptable':
    The Federal Government's Response to Hurricane Katrina
Sinking the Ship of State: The Presidency of George W. Bush
Fool's Gold: The Government's Data Mining Programs
Before the First Snow: Stories from the Revolution

forthcoming:
Collateral Damage in the Marcellus Shale
Photos from the Marcellus Shale

Contents

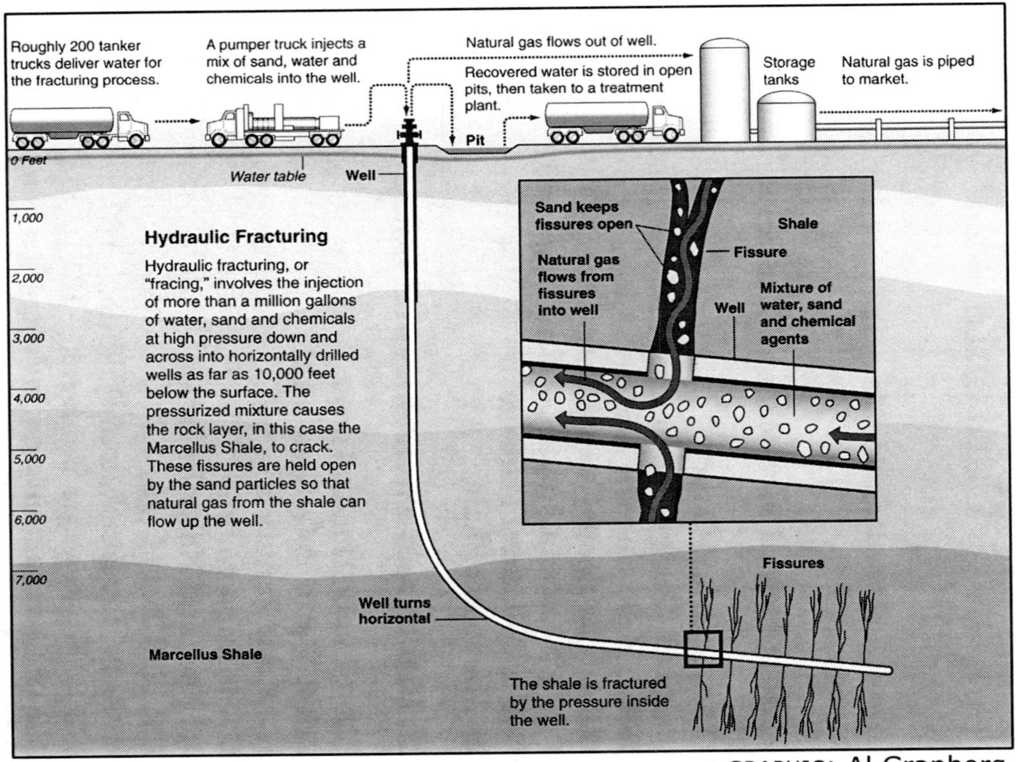

Roughly 200 tanker trucks deliver water for the fracturing process.

A pumper truck injects a mix of sand, water and chemicals into the well.

Natural gas flows out of well.

Recovered water is stored in open pits, then taken to a treatment plant.

Storage tanks

Natural gas is piped to market.

Pit

Well

Water table

0 Feet
1,000
2,000
3,000
4,000
5,000
6,000
7,000

Hydraulic Fracturing

Hydraulic fracturing, or "fracing," involves the injection of more than a million gallons of water, sand and chemicals at high pressure down and across into horizontally drilled wells as far as 10,000 feet below the surface. The pressurized mixture causes the rock layer, in this case the Marcellus Shale, to crack. These fissures are held open by the sand particles so that natural gas from the shale can flow up the well.

Marcellus Shale

Well turns horizontal

The shale is fractured by the pressure inside the well.

Fissures

Sand keeps fissures open

Natural gas flows from fissures into well

Shale

Fissure

Well

Mixture of water, sand and chemical agents

GRAPHIC: Al Granberg

Preface and Acknowledgements

A good friend, David Young of Lewisburg, Pa., an environmental and civil liberties activist, sent me a brief email one day. He asked if I thought there were any First Amendment violations within the proposed Pennsylvania law that regulated the natural gas industry and the gas extraction process known as fracking.

I knew that the proposed law, pushed by the Republican governor and the Republican majorities in both the House and Senate, was controversial. Although seeming to regulate natural gas drilling, it was written by pro-industry lobbyists and legislators.

I didn't want to do much on the issue of fracking. I had been aware of the issue for about three years, and concerned about two years. I was overloaded with other work. I was writing a syndicated newspaper column, doing weekly radio commentaries, and promoting my recently-released book, an historic novel, *Before the First Snow: Stories from the Revolution,* which has an undercurrent of environment and energy activism; the main characters are environmental activists who became involved in a battle in the early 1990s to stop the building of a nuclear waste plant by a corrupt corporation that took safety shortcuts. Not only do the activists have to prove the proposed plant is dangerous to the environment, they have to overcome the lure of jobs in a depressed economy, and show that the temporary boost in economic prosperity is not enough to overcome the decades of health and environmental problems.

I also knew that all the issues and problems in the building and operating of nuclear power plants in the 1980s and early 1990s were almost the same as in the natural gas fracking process two decades later. The government and Big Energy were again collaborating, holding out jobs as the lure to get the people to overlook the hazards of construction and continued operation of one source of energy. Substitute names, and the issues raised decades earlier are the same as today.

Frankly, after an intense time of writing and research to

once again expose problems of nuclear energy, which I acknowledge has significant controls and does employ some of the brightest minds in the energy field although there are significant health risks, I didn't think I had the energy to write about another environmental campaign so soon.

But, once David Young sent me that email, I knew I had to research and write. There was one part of the proposed law that led me to believe there was a bigger story than just a controversial law. Buried deep in the 174-page bill was a section that allowed any natural gas company to withhold disclosing some of the composition of its fracking fluids if the company could establish that it was proprietary information. This would forbid health care professionals from getting the knowledge they might need to treat persons with idiopathic illnesses who came into the emergency rooms or physician offices. If the physician suspected a link between symptoms and the toxic chemicals and compounds in the fracking fluid, the companies would allow the physician to learn the composition of those "trade secrets." But, there was a catch. The physician would have to sign an agreement not to disclose the composition or the cause of illness to anyone, including the patient, medical specialist, or public health worker.

The cavalier attitude of protecting "corporate secrets" at the expense of public health led me to push aside other writing, and begin a month-long investigation, assisted by health care professionals and lawyers.

The article led to many reader responses, but I was still unwilling to do other articles. Frankly, there were many activists who were already protesting, and I didn't know what else I could do to add to the public's knowledge. But, I kept getting reader responses, and suggestions of other stories. After writing two more articles, I finally realized my next book had to be about fracking, a process which most Americans had little knowledge about and which had barely registered on any search engine prior to 2008.

I began this writing project trying to balance all sides, to be objective, to allow the facts to direct my work. But, as I accumulated mounds of evidence, both written and from innumerable interviews, I realized that fracking, even under the best of conditions, is a problem.

One of the first persons I contacted after Pennsylvania made House Bill 1950 law was Dr. Helen Podgainy Bitaxis, who had actively spoken out about the state's gag order. A pediatrician in Coraopolis, Pa., she gave me good information, and left me understanding that those who care about our children also care that no harm comes to them because of the state's aggressive push to accede to corporate wishes that could bring jobs, but at a cost that would impact the health and environment.

I also received excellent advice and information about fracking and its effects from psychologists Diane Siegmund and Kathryn Vennie, and environmental activist Eileen Fay. Dozens of others, including Dr. Wendy Lynne Lee, who are active in the battle for the environment, also assisted, some just by answering a question, others by explaining and answering myriad questions over several months. Among thousands of Pennsylvania activists who have been vigorous in pursuing the truth about fracking are Julie Ann Edgar, Gloria Forouzan, Debbie Lambert, Elizabeth Nordstrom, Liz Rosenbaum, Rebecca Roter, the Rev. Leah Schade, Doug Shields, Nathan Richard Sooy, Michele Novak Thomas, and Maya K. van Rossum, all of whom are active speakers and writers about the effects of fracking.

Long before the establishment media figured out what fracking is and what its impact could be, Abrahm Lustgarten, a *ProPublica* journalist, was researching and writing about the natural gas industry. On top of the story for many years have been *Pittsburgh Post-Gazette* reporters Don Hopey, Laura Olson, Elisabeth Ponsot, and Erich Schwartzel. The AP's Kevin Begos and Mary Esch, and StateImpact/NPR journalists Scott Detrow and Susan Phillips have also been active in digging out the story and accurately reporting it.

Dory Hippauf's "Connecting the Dots" continuing series exposes innumerable connections between government and the energy industry, as well as the industry's shadow and front groups, and connected the "dots" of the mountains of money and influence that went into political campaigns. Karen Feridun's weblog, "Berks Gas Truth," and Iris Marie Bloom's weblog, "Protecting Our Waters," are excellent places to get information from two environmental activists with significant knowledge and a willingness to stand up for what's right.

Judy Morrash Muskauski is the nation's best aggregator of

fracking information. She is the administrator of "Fracking in Northeast Pennsylvania," a website that began as an information site about fracking in Luzerne County, but is now a hub for almost every story about fracking. Her attention to accuracy and the search for information have made it easy to learn about fracking not just in the Marcellus Shale but throughout the country. She and Feridun are also compilers of the largest database of websites and blogs about the fracking industry.

I am especially pleased that Feridun and Muskauski, environmentalists and former librarians, reviewed the manuscript and made numerous suggestions that improved it for publication.

Corey Ellen, MaryJayne Reibsome, and Diana Saavedra of Greeley & Stone Publishers constantly amaze me with their abilities and willingness to make sure that excellence is the only standard acceptable.

Special thanks also goes to my wife, Rosemary R. Brasch, who read the manuscript, commented on it, and made numerous valuable suggestions. As always, I am indebted to my parents, Milton and Helen Haskin Brasch for advice, wisdom, and love.

Working within a department that had become politically charged, most career staff and scientists in the Pennsylvania Department of Environmental Protection responded promptly and accurately to my innumerable questions; off the record, they often gave me information and suggestions that opened new areas of investigation. Similarly, persons working in the natural gas industry also provided assistance, with the understanding I would verify their information and protect their identities.

Fracking Pennsylvania is a fact-based overview of the issues surrounding the natural gas industry and fracking. Although it focuses upon the Marcellus Shale, it looks at cases and issues in other parts of the country. The book is not meant to be a extensive analysis of the science and engineering of the process to extract natural gas nor a comprehensive discussion of the economic, health, environmental, and political issues. It is meant as a basic reference to acquaint people with the issues, with the hope they will dig deeper into areas that directly concern them and rally their friends and neighbors to help protect the health and environment of the people, wildlife, and natural vegetation.

—WALTER M. BRASCH

"The people have a right to clean air, pure water, and to the preservation of the natural, scenic, historic and esthetic values of the environment. Pennsylvania's public natural resources are the common property of all the people, including generations yet to come. As trustee of these resources, the Commonwealth shall conserve and maintain them for the benefit of all the people."

—*The Pennsylvania Constitution (Article I, section 27, 1972)*

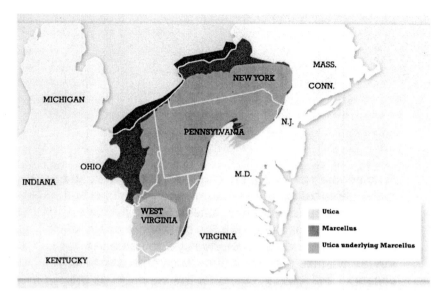

GRAPHIC: Marcellus Shale Coalition

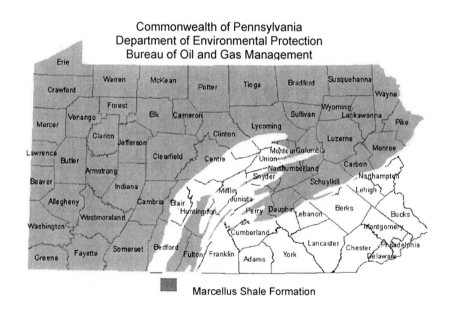

Commonwealth of Pennsylvania
Department of Environmental Protection
Bureau of Oil and Gas Management

Marcellus Shale Formation

Introduction

With the recently-closed Sunoco oil refinery in suburban Philadelphia in the background, a Pennsylvania government official declared his agency "will work hand in glove, very closely, cooperatively and spiritedly with anyone who would want to participate in activities at this facility." [1] Employees, he said, will "work night and day" to help any industry get necessary permits to reopen the facility, perhaps as a natural gas processing refinery. The official continually touted the economic benefits the natural gas industry was bringing to the state.

If that government official was the head of the Department of Community and Economic Development, Labor and Industry, or even the governor's press secretary, it might be just another political speech. But this official was Michael Krancer, and his agency is the Department of Environmental Protection.

On this fifth Friday in June 2012, the public realized it was now official—multi-billion dollar energy companies and the state agency that is charged with protecting the environment were having a public love fest.

The DEP headquarters is located in the Rachel Carson state office building in Harrisburg, named for the woman whose book, *Silent Spring* (1962), was a call for environmental awareness and protection. The irony of an emasculated regulatory agency is not lost upon those who believe energy companies and the government are in collusion.

Almost as an afterthought, Krancer, speaking on behalf of the potential owners of the refinery, claimed they weren't interested "in any short circuiting of environmental protections because they would need to live in this community."

The words sounded good, but the constant noise and odors, products of turning crude oil into gas, were just some of the problems the residents endured in order to benefit from a vigorous job market and the economic benefits that accrued to

1

small businesses, such as cafés and restaurants. The 110 year history that Sunoco Marcus Hook was in the community also included numerous incidents of air and water pollution. The latest, before the plant closed, was toxic air emissions in May, June, and December 2008 that resulted in the DEP, under Gov. Ed Rendell, fining Sunoco $173,000.[2] The Environmental Protection Agency (EPA) had listed the refinery as a regulated site,[3] having had serious violations in the three years prior to its closing in December 2011.

In July, less than two weeks after Krancer said the state was willing to work with energy companies, Gov. Tom Corbett, who took office in January 2011, announced the state awarded a $15 million grant to Brazilian petrochemical company Braskem,[4] which had recently bought a part of the refinery to split propylene, an oil byproduct, into polypropylene, a polymer that is used in dozens of products, including carpets, cups, and chairs.

In the 18 months after Corbett became governor, the natural gas industry, propped up by an industry-favorable law pushed by a Republican-dominated legislature, was creating a temporary economic boom, similar to the booms of the coal and oil industries, both of which led to pollution and worker exploittation before the industries, having stripped the earth, abandoned parts of the state.

Natural gas can be used to heat houses and businesses; provide energy for a number of household appliances, including clothes dryers and ovens; and fuel for transportation. About 54 percent of all single-family residences in the U.S. are heated by natural gas.[5] About 14.8 million vehicles worldwide run on compressed natural gas (CNG);[6] about 114,000 cars, trucks, and buses in the United States have been adapted to run on CNG,[7] which is about half the cost of gasoline derived from oil. The U.S. has about 1,100 fueling stations. Worldwide, about 15 million vehicles use CNG.[8]

The natural gas boom of the past half-decade is primarily because the cost of oil has increased, the U.S. has determined it can not be dependent upon foreign oil, natural gas burns cleaner and is less expensive than other fossil fuels, and the development of fracking, a more economical and efficient method to bring gas to the surface, has been developed.

Natural gas wells and compressor stations have been popping up throughout the Marcellus Shale region; compressor stations prepare natural gas to be pressurized to move through pipes, some extending up to 100 miles, to the next station.

Well-paying jobs have become plentiful; however, most are temporary, ending when the gas companies declare a site no longer profitable. But, hydraulic horizontal fracturing (known as fracking), the process the companies are using to get to the gas more than a mile beneath the surface, is leaving in its wake health and environmental issues that could be as serious as those that surrounded the coal and oil industries.

President Barack Obama has repeatedly spoken against the heavy use and dependence upon fossil fuels, and sees the use of natural gas as a transition fuel to expanded use of wind, solar, and water energy. In his January 2012 State of the Union, the President advocated the development of natural gas exploration, which "will create jobs and power trucks and factories that are cleaner and cheaper, proving that we don't have to choose between our environment and our economy."[9]

There is also another consideration. "Geopolitics play a significant role in whether a number of gas projects are realized and come online and where pipelines are built. As each scenario demonstrates, individual country decisions about natural gas resources can have dramatic impacts on responses in international discourse," according to a July 2012 research analysis published by the Kennedy School of Government at Harvard University.[10] Amy Myers Jaffe and Dr. Meghan L. O'Sullivan, co-editors of the study, also point out, "The relative fortunes of the United States, Russia and China—and their ability to exert influence in the world—are tied in no small measure to global gas developments and vice versa."[11] Gazprom, Russia's state-owned energy corporation, and the world's largest exporter of natural gas, had a $44 billion profit in 2011.[12] However, two factors may diminish Gazprom's global dominance—natural gas in the U.S. was selling for about $3 a unit domestically while Gazprom was exporting natural gas at $10 a unit and, according to the AP, "Gazprom halted plans [in September 2012] to develop a new arctic gas field, saying it couldn't justify the investment now."[13]

3

What is Fracking?

High volume horizontal hydraulic fracturing, a combination of horizontal drilling and hydraulic fracturing, commonly known as fracking, is the controversial method of forcing five to nine million gallons of water, propping agents (usually sand or ceramic beads), and as much as 100,000 gallons of gases and chemicals deep into the earth. With the water and propping agents, chemicals and radioactive isotopes at a pressure of up to 15,000 pounds per square inch are forced into a rock formation as deep as 12,000 feet below the earth's surface to open channels and force out natural gas and fossil fuels. The additives "are used to prevent pipe corrosion, kill bacteria, and assist in forcing the water and sand down-hole to fracture the targeted formation," explains Thomas J. Pyle, president of the Institute for Energy Research.

Natural gas companies place concentric steel rings (casings) within the well bore and then pour concrete into the bore to try to keep toxic chemicals deep in the earth from polluting the water sources they must drill through to get to the gas, and to reduce subsequent air and water pollution. For the first few weeks, while the companies drill through the shale, tall derricks identify the locations; after that, the derricks come down; the wells remain, now connected to pipelines and compressor stations.

"The gas industry tells the public it's safe because they're extracting the gas so far below the ground surface that any contaminants they use will never make their way up to an aquifer," Dr. Patricia Culligan, professor of civil engineering and engineering mechanics at Columbia University, told *Columbia Magazine*.[14] That is only partially correct. According to Dr. Culligan, "[T]hat assumes that the gas well is properly developed and sealed. If it's not, contaminants have a pathway to migrate into upper aquifers. And once you've polluted an aquifer, it's almost impossible to undo the damage." Well casings in Pennsylvania during 2010 and 2011 had a failure rate of 6.2 percent, leading to pollution of ground water,[15] according to an analysis of DEP data by Dr. Anthony Ingraffea, professor of civil and structural engineering at Cornell University. "No matter what the industry tells you, their own data proves conclusively to any reasonable scientist or engineer that it is impossible to design any well so it will never leak," says Dr. Ingraffea.[16]

After drilling down vertically, the natural gas company creates a perforated lateral borehole, 90 degrees from the vertical hole, which fractures the shale and rock for up to about 6,000 feet.[17] The shale itself is between 40[18] and 900[19] feet thick; natural gas in the shale is trapped between impervious layers of limestone. "Fracking turns solid bedrock into broken shards whose cracks become potential pathways for contamination, some of it radioactive. Broken shale is not reparable by any known technology," says Dr. Sandra Steingraber, a biologist and distinguished scholar in residence at Ithaca College.[20]

The mined gas is transported by truck and by more than 1.5 million miles of pipelines.[21] Extensive truck traffic to bring water, fracking fluids, and equipment to the drilling sites and take gas and wastewater away have clogged public roads and highways. More important, says environmentalist Robert F. Kennedy Jr., "the Industry now acknowledges that it absolutely cannot afford to pay localities the costs of roads damaged from the thousands of truck trips per wellhead, leaving those ruinous costs to local taxpayers."[22]

About half of all water and various elements and chemicals—known variously as wastewater, flowback, blowback, or brine, saltier than seawater—is brought to the surface and must be disposed. In most drilling operations, this flowback is captured in storage tanks or ponds, where it is pumped into tanker trucks for disposal at sewage facilities or reused in other drilling operations. In some cases, it just lies in unprotected open ponds. The other half of the mixture stays within the ground, sometimes affecting water supplies; however, that half, contaminated with fracking chemicals, will eventually come to the surface.

A 2011 analysis showed that companies were now using an average of 2.4–7.8 million gallons of fluid per non-conventional (fracked) well;[23] the amount is based upon the geology of the basin. Pennsylvania energy companies use an average of 4.3 million gallons of water per frack,[24] according to data compiled by FracFocus and analyzed by EcoWatch. Between January 2011 and August 2012, companies drilling in Pennsylvania used about

8.2 billion gallons of water; overall, energy companies through-out the U.S. used about 65.9 billion gallons of water during that time frame.

Water is so critical to the process that during the drought in the Midwest in 2012, oil and gas companies were buying water from farmers and municipalities and trucking in water from as far away as Pennsylvania.[25] Agriculture fields and, sometimes, livestock suffered because the industry needed water, and were paying premium prices, as much as $1,000–$2,000 for about 326,000 gallons (an acre foot);[26] the normal price is about $30–$100 for the same amount. Energy companies are "going to pay what they need to pay," said Dr. Reagan Waskom, director of the Colorado Water Institute at Colorado State University.[27] In most cases, the natural gas industry was able to outbid farmers for water supplies.

A newer significantly more expensive technique is being developed. Propane Hydraulic Fracturing uses large volumes of liquid propane and smaller quantities of butane mixed with phosphoric acid to create a gel, which is then mixed with sand and several compounds. Gary P. Hoffman, of Applied Thermo-dynamics, explains:

> "This gelled hydraulic fluid is then forced down the per-forated well bore exactly in the same manner as the slick water [low viscosity water, with less sand and chemicals] is under the conventional method. Once the rock is fractured, the high pressure is released. This causes the gel to flash back into gaseous propane. The propane gas, and the methane gas, now come roaring back up the well bore as a mixed, high explosive gas. Here they are controlled, it is hoped, by a blow-out preventer, and are gradually separated into the two gasses."[28]

Although it is less of a problem than oil or coal emissions, about 78 percent of unprocessed natural gas is methane,[29] one of the major contributors to ozone layer depletion. "[W]hen released directly into the atmosphere [methane] is a potent greenhouse gas—more than 20 times more potent than carbon dioxide," says Lisa Jackson, Environmental Protection Agency administrator.[30] The other major problem with methane is that it can migrate into water supplies.[31]

The natural gas industry defends fracking as safe and efficient. Thomas J. Pyle, claims fracking has been "a widely deployed as safe extraction technique,"[32] dating back to 1949 when the first commercial well was drilled by Halliburton. What he doesn't say is that energy companies until recently had used low-pressure methods to extract natural gas from fields closer to the surface than the current high-pressure high-volume technology that extracts more gas but uses significantly more water, chemicals, sand, and elements. An investigation by New York state in 1998 revealed that natural gas companies used only 20,000–80,000 gallons of water per conventional well.[33]

Because of the chemicals added to the water, the methane that can't be extracted, and the chemicals and elements released from the earth during fracking, the water can't be used again, except to frack other wells.

Advocates of fracking argue natural gas is "greener" than coal and oil energy, with significantly fewer carbon, nitrogen, and sulfur emissions. "Over its full cycle of production, distribution, and use, natural gas emits just over half as many greenhouse gas emissions as coal does for equivalent energy output," according to the Worldwatch Institute.[34] But, there are other realities. First, escaped methane from natural gas into the air and water increase problems with public health and the environment; second, production and distribution errors and problems mining natural gas make it the equivalent of coal and oil.

Each of Pennsylvania's 8,568 active wells, as of the beginning of July 2012,[35] takes up about 8.8 acres;[36] access roads, pipelines, and staging areas take up an additional 21 acres.[37]

Oil and gas companies drilled about 245,000 conventional wells between 1991 and 2000;[38] between 2001 and 2010, the companies drilled about 405,000 wells. As of January 1, 2012, there were about 504,000 active wells in 31 states;[39] natural gas companies used fracking procedures in about 90 percent of the wells.[40] While vertical fracking had been used for decades, it was horizontal fracturing that changed the industry.

At the beginning of the 21st century, deep shale produced only about one percent of all natural gas; however, within a decade, it was producing about 30 percent.[41] About 45 percent of all natural gas produced in America is expected to be produced by fracking and horizontal drilling by 2035.[42]

ORIGINS OF FRACKING

High-volume horizontal fracking was first developed in the Barnett Formation in Texas. Dan Steward, geologist and former vice-president of Mitchell Energy, the company that explored and then developed natural gas fracking, was one of the pioneers of the process that even the harshest critics acknowedge is the result of creative and brilliant engineering. "Some people thought it [fracking] was stupid," Steward told the Associated Press in September 2012; even in his own company, "probably 90 percent of the people" didn't believe fracking could be possible or, if so, would not be profitable.[43] In an interview with the Breakthrough Institute, a liberal think tank with a focus on global ecological problems, Steward discussed the origins of fracking:

"In the seventies we started looking at running out of gas, and that's when the DOE started looking for more.

"The [Department of Energy's] Eastern Gas Shales Project [in the Appalachia basin in 1976] determined there was a hell of a lot of gas in shales. It was the biggest accumulation of data and knowledge to date. It set the stage for people to have the basic background and caused people to start asking questions, and that's always important.

"They did a hell of a lot of work, and I can't give them enough credit for that. DOE started it, and other people took the ball and ran with it. You cannot diminish DOE's involvement. . . .

"Mitchell got involved in the Cotton Valley limestone looking for gas, but it was tight rock, and George [Mitchell] said, 'I want to frack it.' But he had a hard time to get his people to go along.

"Mitchell was interested in Barnett and his geophysicist said, 'It looks similar to the Devonian, and the government's already done all this work on the Devonian.' . . .

"In the 1990s they [the Department of Energy] helped us to evaluate how much gas was there, and evaluate the critical properties as compared to Devonian shale of Appalachia basin. They helped us with our first horizontal well. They helped us with pressure build-ups. And we worked with them on crack mapping. In 1999 we started working with GTI (formerly GRI) on re-fracks of shale wells. . . .

"[At] the time we started trying the Barnett, the thinking

was we had to have open natural fractures. And so as we moved along we drilled wells and built the database.

"There was trial and error. Frequently that's what has to happen. You have to take best science and trial and error things. That's how Barnett [Shale] got started. . . .

"We tried on two wells one time and a third time three years later where we were trying this microseismic, and we thought when they get the bugs worked out this is going to be break. Until 1997 the bugs hadn't been worked out. . . .

"By the year 2000, Mitchell Energy had proven shale as a workable and viable. The energy industry recognized it, but financial markets didn't recognize until 2002, and politicians only realized it in 2006.

"In 2002 we actually started drilling horizontals in the Barnett as part of Devon Energy, which had bought Mitchell. They started with single fracks in the horizontals. And then from frack mapping they realized that we're not connecting with much of the rock. . . .

"It wasn't until 2006 and 2007 that we finally found the porosity, thanks to work done by the Bureau of Economic Geology, which I think gets state and federal funds. We knew there was porosity in the rock but couldn't see it with available technology. They experimented with ways to look at the rock at a higher resolution. And applied argon milling used in metallurgy where you could mill a surface on a piece of metal that was so clean that you could look at surface with high resolution."[44]

The AP's Kevin Begos, interviewing Alex Crawley, former associate director of the Department of Energy's National Petroleum Technology Office, reported:

"The work wasn't all industry or all government, but both.

"One step at a time, the problems of shale drilling were solved. Crawley said Energy Department researchers processed drilling data on supercomputers at a federal lab. Later, technology created to track sounds of Russian sub-marines during the Cold War was repurposed to help the industry use sound to get a 3-D picture of shale deposits and track exactly where a drill bit was, thousands of feet underground.

"'It was a lot of pieces of technology that the industry thought would help them. Some worked out, some didn't,' Crawley said."[45]

AVAILABLE NATURAL GAS

The U.S. Energy Information Administration (U.S. EIA) estimates there may be as much as 112 trillion cubic meters of natural gas, about 3,950 trillion cubic feet, available to be mined worldwide.[46] (One cubic meter equals about 35.3 cubic feet). Potential natural gas in China, which is at the verge of developing the technology, may be about 36.1 trillion cubic meters.

There are varying estimates of the natural gas in shales in the United States, with estimates by the U.S. EIA of 410 trillion cubic feet[47] to as much as 2,000 trillion cubic feet,[48] about half of all available natural gas worldwide. However, estimates closer to 2,000 trillion cubic feet often come from the industry itself "and include every bit of natural gas known or imagined, whether or not is economically or technologically viable to recover," says Karen Feridun, founder of Berks Gas Truth and one of Pennsylvania's leading environmental activists. Nevertheless, natural gas could temporarily replace coal and oil, and serve as a transition to wind, solar, and water as primary energy sources, releasing the United States from dependency upon fossil fuel energy and allowing it to be more self-sufficient.

There are several shales in the United States which contain natural gas—Barnett and Barnett–Woodford shales in Texas; Bakken shale in North Dakota and Saskatchewan; Fayetteville shale in Arkansas; Haynesville and Eagle Ford shales in the Louisiana area; Mancos and Lewis shales in the San Juan Basin, primarily in New Mexico, Colorado, and parts of Utah; the Monterey Shale in California's San Joaquin Basin, which yields primarily oil and some gas through conventional drilling methods; the South Georgia Basin in South Carolina, Georgia, and northern Florida; the South Newark Basin, which extends through parts of New Jersey and southern Pennsylvania; the Utica Shale, a deeper shale which ranges from parts of Ontario and Quebec in Canada, through parts of New York, Pennsylvania, Ohio, and West Virginia; the Woodford Shale in Oklahoma; several smaller shales along the East Coast from Delaware into South Carolina; and the Marcellus Shale, about 95,000 square miles.

The Marcellus Shale—which extends beneath the Allegheny Plateau, through southern New York, much of Pennsylvania, east Ohio, West Virginia, and parts of Maryland and Virginia—lies on top of the Devonian and Utica shales. The Marcellus Shale was created during the Middle Devonian epoch, about 400 million years ago. The Marcellus Shale, also known as the Marcellus Formation, was named by geologist James Hall in 1839 for Marcellus, N.Y., a small village near Syracuse, where an outcrop of the shale was first discovered.[49] (An outcrop is a geological formation that is usually covered by soil and vegetation but is exposed by erosion.)

The Marcellus Shale is one of the nation's largest sources for natural gas mining. In 2002, the U.S. Geological Survey (USGS) estimated there were about 1.9 trillion cubic feet of natural gas in the Marcellus Shale,[50] not substantial enough to allow profitable extraction by energy companies. However, the following year, Range Resources drilled the first well in Washington County, Pa., and other companies soon drilled exploratory wells. The gas rush boom began about 2008 when Dr. Terry Engelder, professor of geosciences at Penn State, figured there might be about 500 trillion cubic feet of gas in the shale. That figure might be inflated. The USGS revised figure is about 84 trillion cubic feet of available natural gas.[51] The Marcellus Shale may contain only about 141 trillion cubic feet of natural gas, according to the U.S. Energy Information Administration.[52] Other estimates vary, but none are even half that provided by Dr. Engelder.

During 2012, the Marcellus shale yielded about seven billion cubic feet of gas a day, about one-fourth of the nation's gas production.[53] However, contrary to some industry claims that the natural gas could provide energy for more than a century, the reality is that it may provide only enough for a couple of decades. The high-end estimates do not take into consideration that at present only 10–15 percent of all natural gas in the Marcellus Shale is viable; the costs to go after the rest of the available gas may not be financially justified if gas prices to consumers have to increase significantly, says Karen Feridun of Berks Gas Truth.

Underground Storage

The Finger Lakes is a series of 11 long and narrow lakes between Ithaca and Rochester, N.Y. The region is one of the nation's leading tourism areas, with extensive water recreation, waterfalls, hiking trails, museums, art and photo galleries, colleges, farmer's markets, boutiques, and outdoor malls. Almost 100 wineries and vineyards, all taking advantage of the "lake effect" and good soil, make the area the largest wine growing region in New York.

It's also where Inergy, one of the nation's largest independent gas storage and transportation companies, paid U.S. Salt about $65 million in August 2008 to acquire 576 acres on the southwestern side of Seneca Lake, the largest of the Finger Lakes, and plans to spend an additional $40–50 million to convert abandoned salt caverns into what may become the largest gas storage facility in northeastern United States.[54] It's also where Dr. Joseph Campbell and Yvonne Taylor organized Gas Free Seneca, a grassroots organization of several hundred residents, more than 150 businesses, and more than 5,000 persons who signed a petition to state regulatory agencies asking that they block Inergy's plans.

Inergy plans to store more than 600,000 barrels of butane in Well 58, and as many as 1.5 million barrels of propane in a series of four other caverns, known collectively as Gallery 1. Inergy, says a fact sheet posted on Gas Free Seneca, "has documented plans to increase their salt cavern storage capacity to five million barrels (210 million gallons) of LPG and has recently acquired . . . two billion cubic feet of underground natural gas storage [from the New York State Electric & Gas Co.] with plans to expand to 5–10 billion cubic feet."[55] The company stores the mined gas for companies that do the drilling. The caverns will have more product in the warm months that coincide with the tourist season.

"This type of industrialization will destroy the local, sustainable economy that brings several hundred million dollars into New York on an annual basis," says Taylor, who points out that the storage of natural gas liquids (NGLs) will bring active burning flare stacks, a six-track train depot, a truck depot (which would add an additional eight trucks to narrow roads per hour), open brine pits, compressor stations, and pipelines."

Trains carrying liquid propane and liquid butane would cross a 75-year-old train trestle that spans the Watkins Glen Gorge. Because propane and butane are heavier than air, "if there were an accident and one of the cars breached, the gas would form a dense, explosive vapor cloud and funnel right down to the gorge and into the heart of Watkins Glen" where it could ignite merely by someone lighting a cigarette, says Dr. Campbell.

Dr. Campbell has sailed on the lake about a dozen years, "and I've never seen this kind of activity; drilling went on non-stop for about 15 months." And that drilling was solely to test the caverns and upgrade them to accept natural gas and NGLs liquids when the state issues its permits. If past practices are any indication, Inergy will probably receive property tax benefits. In other places where it has built, Inergy has often violated state and local regulations and has freely used the power of state-approved eminent domain rules to seize private property.

But, more important than the possible destruction of small business economy is the possibility that the storage facility will bring with it environmental and public heath issues. Within a three week period in July 2012, says Taylor, "Inergy has had two equipment failures resulting in brine spills which killed trees and vegetation along the hillside." The company also received permission from the state to discharge 44,000 pounds of chloride into Seneca Lake per day. That lake provides drinking water to more than 100,000 individuals.

Gas Free Seneca has raised objections that there are two major problems that need to be addressed: potential cavern collapse and potential cavern leakage through faults in walls, roofs, and the floor. It is a concern that the state's Department of Environmental Conservation is also looking into, especially since the previous owner abandoned the salt cavern after the roof collapsed in an earthquake. But, the facts of that roof collapse and other issues of the integrity of the caverns are protected as a "trade secret," a ruling issued by the DEC in November 2010. "There's a lot of information about these caves that we can't get our hands on," says Dr. Campbell.

In a statement to DC Bureau, an independent news service, Taylor, said she believes "The DEC and the EPA are not in place to protect the environment or care for individuals and small businesses so much as they are there to promote heavy industry and protect large corporations."[56]

Neither Joseph Campbell nor Yvonne Taylor are long-time activists. Dr. Campbell, a chiropractor, says he had a "limited role" in successfully opposing Chesapeake Energy from creating a wastewater disposal injection well near Keuka Lake, but the probable destruction of a scenic area for corporate profit caused him to become more active. Taylor, a speech and language teacher, was previously active in stopping gas drilling in the Finger Lakes National Forest on the east shore of Seneca Lake.

The U.S. currently has about 3.7 trillion cubic feet of natural gas in about 400 underground storage sites.[57] About 80 percent of all underground storage is in depleted gas or oil fields; the rest are in aquifers or salt caverns. However, in the past two decades, the number of salt cavern storage sites has grown steadily, mostly in the Gulf Coast areas.[58]

A sinkhole near Bayou Corne, La., about 80 miles west of New Orleans, is one of the reasons the residents of the Finger Lakes may be worried. The walls of an underground three square mile salt cavern,[59] which Texas Brine had been mining for the fracking industry until 2011, collapsed in August 2012. The U.S. Geological Survey determined that the mining probably caused several micro-earthquakes that had weakened the walls.[60] The collapse and release of methane gas forced the state to issue a mandatory evacuation for about 300 residents living in about 150 homes. The depth of the sinkhole was originally 449 feet at its deepest part.[61] Although the depth decreased because of collapsing walls and earthslides, the width continued to expand; by the end of 2012 the area of the sinkhole was more than eight acres at the surface.[62] To rid the area of the gas and its escape into local aquifers and swamps, Texas Brine built several vents to burn off the escaping gas, leading to additional air pollution.[63] Among other health problems are the presence of the toxic benzene and diesel fuels in the release of oil.[64]

Pennsylvania has about 673.5 billion cubic feet stored in 65 sites.[65]

Bans on Horizontal Fracking

An extensive study of fracking commissioned by the European Union revealed, "Risks of surface and ground water contamination, water resource depletion, air and noise emissions, land take, disturbance to biodiversity and impacts related to traffic are deemed to be high in the case of cumulative projects."[66] The 292-page study recommended that no fracking be allowed near areas where water is used for drinking.

Because of significant questions about health and pollution issues related to fracking, several countries have banned the use of fracking. Bulgaria, France, Germany, Luxembourg, and Ireland have banned all fracking operations. The Czech government is seriously considering a ban. Fracking in several other European countries, however, may be a moot issue. Test wells in several countries reveal that extracting the gas may not be commercially feasible.

Almost two-thirds of Canadians oppose fracking, according to a poll conducted in January 2012 by Environics Research.[67] Quebec, British Columbia, New Brunswick, and Nova Scotia have also banned fracking pending full studies.

In April 2011, South Africa banned Shell Oil from using fracking to extract natural gas in the Karoo Desert.[68] [F]racking fluid *will* contaminate the groundwater. There is not doubt at all," said Dr. Gerrit van Tonder of South Africa's Institute for Groundwater Studies at University of the Free State.[69] However, in September 2012, the South African government lifted that moratorium on exploration following a study that correlated safe extraction with the reality that the country has about 485 trillion cubic feet of natural gas, most of it in the Karoo Desert, and drilling would boost economic recovery and lower oil dependency.[70]

Several U.S. states have either banned fracking or placed moratoriums on it until full health impact analyses could be completed. Both New Jersey and New York placed moratoriums on well permits while they evaluate the health and environmental risks. The New Jersey legislature also banned disposal of fracking wastewater, targeting the possibility that companies operating in Pennsylvania would transport much of the wastewater from the eastern part of the state to New Jersey. However, Gov. Chris Christie vetoed it.

The New York moratorium began in 2008; the ban was extended until July 1, 2011, following majority votes in the state senate and assembly. Signing the legislation in December 2010, Gov. David Patterson said New York "would not risk public safety or water quality."[71] The ban was continued under the administration of Gov. Andrew Cuomo.

Vermont banned fracking, but the action by the legislature and Gov. Peter Shumlin is symbolic since Vermont has no deep earth natural gas deposits. Nevertheless, the natural gas Industry lobbied against the proposed legislation and then spoke out after the legislature passed it. Rolf Hanson of the American Petroleum Institute said the Legislature's action "follows an irresponsible path that ignores three major needs: jobs, government revenue and energy security."[72]

Gov. Bev Perdue (D-N.C.), who supports fracking and natural gas development, vetoed a Republican-sponsored bill in July 2012 that would have removed that state's ban on fracking. "Our drinking water and the health and safety of North Carolina's families are too important [and] we can't put them in jeopardy by rushing to allow fracking without proper safeguards,"[73] said Gov. Perdue in announcing her veto.

Gov. Martin O'Malley (D-Md.) issued an executive order to delay drilling in Maryland until a special commission determined fracking could be conducted safely. However, oil and gas energy lobbyists influenced enough members of the legislature that it has refused to fund that study. A resolution endorsed by several Maryland environmental, political, and civic organizations urges that "This statutory moratorium should only be permitted to expire if and when detailed and transparent studies prove that fracking activities will not cause harm to our public health, rural communities, natural environment, and global climate." Gov. O'Malley's moratorium expires in January 2015, the end of his term.

Dozens of cities throughout the United States have enacted bans or moratoriums on natural gas drilling within their city limits, usually citing health and environmental concerns for their actions.

New York City asked the state to ban all natural gas drilling in the watershed. Fracking presents "unacceptable threats to the unfiltered fresh water supply of nine million New Yorkers," said Steven Lawitts, commissioner of the city's Department of Environmental Protection, in December 2009. Mayor Michael Bloomberg agreed; a representative of the mayor told Reuters, "Based on all the facts, the risks are too great and drilling simply cannot be permitted in the watershed."[74]

By a 9–0 vote of the city council in November 2010, Pittsburgh became the first Pennsylvania city to ban natural gas drilling. The Council cited health concerns as its reason to ban drilling.[75] City Councilman Doug Shields called the natural gas industry "arrogant," and said it placed profits ahead of citizen health and the environment. Among the leases that Huntley & Huntley signed were for 1,060 acres of land beneath 15 cemeteries under jurisdiction of the Catholic Cemeteries Association of the Diocese of Pittsburgh.[76] Council president Darlene Harris, responding to industry claims it would bring jobs to the region, was brutally honest—"They're bringing jobs all right. There's going to be a lot of jobs for funeral homes and hospitals. That's where the jobs are. Is it worth it?"

Two months after the Pittsburgh city council voted to ban fracking, the city council of Philadelphia, the state's largest city, voted to ban all fracking in the Delaware River Basin.[77]

In a letter to President Barack Obama, Physicians, Scientists, & Engineers for Healthy Energy (PSE), argued:

> "There is a growing body of evidence that unconventional natural gas extraction from shale (also known as 'fracking') may be associated with adverse health risks through exposure to polluted air, water, and soil. Public health researchers and medical professionals question the continuation of current levels of fracking without a full scientific understanding of the health implications. . . .
>
> "There is a need for much more scientific and epidemiologic information about the potential for harm from fracking. To facilitate a rapid increase in fracking in the United States without credible science is irresponsible and could potentially cause undue harm to many Americans."[78]

While other cities, states, and countries are either banning fracking or suspending fracking until a full health and environmental analysis can be completed, Pennsylvania under the Tom Corbett Administration is "handing out permits almost like popcorn in a theater," says Diane Siegmund, a psychologist from Towanda, the county seat of Bradford County, where more than 1,200 gas wells (conventional and non-conventional) are located.[79]

Pennsylvania, at the epicenter of the Marcellus Shale exploration, rushed to embrace the natural gas industry and its use of fracking, apparently disregarding its own Constitution that requires the state to recognize "The people have a right to clean air, pure water, and to the preservation of the natural, scenic, historic and esthetic values of the environment."[80]

Like a five-year-old running from house to house on Halloween to gather as much candy as possible, Tom Corbett and the Republicans have been wide-eyed ecstatic about natural gas drilling. Between Jan. 1, 2005, and the end of 2012, the Pennsylvania Department of Environmental Protection issued 12,406 permits to 124 companies,[81] and denied only 48 requests to construct unconventional gas wells.[82]

Photo: Jacques-Jean Tiziou
(courtesy of Protecting Our Waters)

Protesters at the Shale Gas Outrage,
September 2012, in Philadelphia.

PART I:
Historical, Political, and Economic Issues

The forests that once dominated Pennsylvania were destroyed during the lumber boom era of the second half of the nineteenth century.

PHOTO: Department of Energy

The Shippingport Atomic Power Station, about 35 miles northwest of Pittsburgh, was the nation's first nuclear power plant developed solely for peaceful use.

CHAPTER 1
A Brief History of America's Energy Policies

The history of energy exploration, mining, and delivery is best understood in a range from benevolent exploitation to worker and public oppression. A company comes into an area, leases or buys land in rural and agricultural areas for mineral rights, increases employment, usually in a depressed economy, strips the land of its resources, creates health problems for its workers and those in the immediate area, and then leaves.

It makes no difference if it's timber, oil, coal, or nuclear. All energy sources are developed to move mankind into a new era; all energy sources are developed to bring as much profit to corporations as quickly as possible, often by exploiting the workers.

Before the settlement of Pennsylvania in the 1680s, more than 20 million acres of forests covered almost all of the land. During the latter half of the nineteenth century, the lumber industry had clear-cut several million acres, leading Pennsylvania into an era that rivaled even the Gold Rush in California. By World War I, the companies had stripped the land, taken their profit, and then moved on, leaving devastation in their wake. Only when the people finally realized that destroying the forests led to widespread erosion and flooding did they begin to reforest the state. Almost a century after the lumber companies denuded the forests, the natural gas industry, with encouragement from the state, have leased more than 150,000 acres of forests for wells, pipelines, and roads.[83]

Between 1859, when an economical method to drill for oil was developed near Titusville, Pa., and 1933, the beginning of Franklin D. Roosevelt's "New Deal," Pennsylvania, under almost continual Republican administration, was among the nation's most corrupt states.[84] The robber barons of the timber, oil, coal, steel, and transportation industries, enjoying and contributing to the Industrial Age of the 19th century, essentially bought their right to be unregulated. In addition to widespread bribery, the energy industries, especially coal, assured the election of preferred candidates by giving pre-marked ballots to workers, many of whom were immigrants and couldn't read English.

When the coal companies determined underground mining was no longer profitable, they began strip mining, shearing the tops of hills and mountains to expose coal, causing environmental damage that could never be repaired even by the most aggressive reforestation program. Pennsylvania is the only state producing anthracite coal, and is fifth in the nation in production of all coal, behind Wyoming, Kentucky, West Virginia, and Texas.[85]

John Wilmer, an attorney who formerly worked in the Pennsylvania Department of Environmental Protection (DEP), in a letter to the editor of *The New York Times* in March 2011, explained that "Pennsylvania's shameful legacy of corruption and mismanagement caused 2,500 miles of streams to be totally dead from acid mine drainage; left many miles of scarred landscape; enriched the coal barons; and impoverished the local citizens."[86] His words are a warning about what is happening in the natural gas fields.

Every method of extracting energy from the earth yields death and injury to the workers and residents. More than 100,000 coal miners were killed, often from structural failures within the mines, gas poisonings, explosions, and roof collapse. Long-term catastrophic effects from mining also include pneumoconiosis, also known as Black Lung Disease, the result of the inhalation of coal dust within the mines. Worker and resident protection often don't occur until decades after a new energy source is mined. For coal mining, although there were several protections brought about by the United Mine Workers, it wasn't until 1969 when the Federal Coal Mine Health and Safety Act became law that health and environmental protection advanced. Congress improved the Act in 1977 and 2006.

The nation's first commercial nuclear power plant to develop peaceful uses of energy was the Shippingport Atomic Power Station, along the Ohio River in Beaver County, Pa., about 35 miles northwest of Pittsburgh. The plant went online in December 1957 and stayed in production through October 1982. During the last four decades of the twentieth century, the nation built 132 nuclear plants, with politicians and Industry claiming nuclear energy was clean, safe, efficient, and would lessen the nation's ties to oil. Chernobyl, Three Mile Island, Fukushima Daiichi, and thousands of violations issued by the Nuclear Regulatory Agency, have shown that even with strict operating guidelines, nuclear energy isn't as clean, safe, and as efficient as claimed. Like all other energy industries, nuclear power isn't infinite. Most plants have a 40–50 year life cycle. After that, the plant becomes so radioactive that it must be sealed. Pennsylvania is second in the nation, behind Illinois, in production of electricity from nuclear reactors.[87]

In the early 21st century, the natural gas industry follows the model of the other energy corporations, and uses the same rhetoric. The Heartland Institute, a think tank which says it exists to "promote free-market solutions to social and economic problems, claims, "Shale extraction has proven remarkably safe for the environment and the newfound abundance of domestic natural gas reserves promises unprecedented energy prosperity and security."[88]

Much of the development costs for fracking were funded by the U.S. Department of Energy. Beginning in 1975, the federal government encouraged development of natural gas, and in three decades contributed about $137 million for natural gas research.[89]

The Resource Conservation and Recovery Act (RCRA),[90] passed in 1976 during the Gerald Ford administration (1974–1977), established a strict protocol for hazardous waste management. In 1980, Congress passed two major bills to specifically benefit the the development of natural gas Industry. First, it established significant tax breaks to encourage drilling of non-conventional wells; second, it passed the Comprehensive Environmental Response, Compensation, and Liability Act (CERCLA)[91] that created the "superfund" for cleaning hazardous waste, and held the energy industry financially accountable for hazardous waste that could affect health. However, Congress specifically exempted oil and natural gas industries from CERCLA.

In 1988, the last year of the Reagan–Bush administration,

23

Congress exempted oil and gas hazardous waste from oversight by the RCRA.

A *New York Times* article in March 2011 summarized what had happened during the Reagan years:

> "When Congress considered whether to regulate more closely the handling of wastes from oil and gas drilling in the 1980s, it turned to the Environmental Protection Agency to research the matter. E.P.A. researchers concluded that some of the drillers' waste was hazardous and should be tightly controlled.
>
> "But that is not what Congress heard. Some of the recommendations concerning oil and gas waste were eliminated in the final report handed to lawmakers in 1987. 'It was like the science didn't matter,' Carla Greathouse, the author of the study, said in a recent interview. 'The industry was going to get what it wanted, and we were not supposed to stand in the way.'
>
> "E.P.A. officials told her, she said, that her findings were altered because of pressure from the Office of Legal Counsel of the White House under Ronald Reagan. A spokesman for the E.P.A. declined to comment."[92]

Under the Clinton Administration (1993–2001), "national environmental targets were made more stringent, and environmental quality improved," according to an 84-page report by Sheila M. Cavanagh, Robert W. Hahn, and Robert N. Stavins, and published in September 2001.[93] According to the authors:

> "Most important among the new targets were the National Ambient Air Quality Standards (NAAQS) for ambient ozone and particulate matter, issued by EPA in July 1997, which could turn out to be one of the Clinton Administration's most enduring environmental legacies. Also, natural resource policy during the Clinton years was heavily weighted toward environmental protection. Environmental quality improved overall during the decade, continuing a trend that began in the 1970s, although improvements were much less than during the previous two decades.
>
> "Second, the use of benefit-cost analysis for assessing environmental regulation was controversial in the Clinton Administration, while economic efficiency emerged as a central goal of the regulatory reform movement in the Congress during

24

the 1990s. When attention was given to increased efficiency, the locus of that attention during the Clinton years was the Congress in the case of environmental policies and the Administration in the case of natural resource policies. Ironically, the increased attention given to benefit-cost analysis may not have had a marked effect on the economic efficiency of environmental regulations.

"Third, cost-effectiveness achieved a much more prominent position in public discourse regarding environmental policy during the 1990s. From the Bush Administration through the Clinton Administration, interest and activity regarding market-based instruments for environmental protection—particularly tradeable permit systems—continued to increase.

"Fourth, the Clinton Administration put much greater emphasis than previous administrations on expanding the role of environmental information disclosure and voluntary programs. While such programs can provide cost-effective ways of reaching environmental policy goals, little is known about their actual costs or effectiveness.

"Fifth and finally, the Environmental Protection Agency placed much less emphasis on economic analysis during the 1990s. EPA leadership was more hostile to economic analysis than it had been under the prior Bush Administration, and it made organizational changes to reflect this change in priorities."[94]

The energy policy during the eight years of the George W. Bush–Dick Cheney Administration (2001–2009) was to give favored status to the industry, often at the expense of the environment. In addition to negating Bill Clinton's support for the Kyoto Protocol to reduce greenhouse-gas emissions, an act signed by 191 countries, former oil company executives Bush and Cheney pushed to open significant federal land to drilling. Included in the proposal was the 19 million acre Arctic National Wildlife Refuge (ANWR); drilling in the ANWR would disrupt the ecological balance in one of the nation's most pristine areas.

A study by the EPA, published in 2004, concluded that fracking was of little or no risk to human health.[95] However, Wes Wilson, a 30-year EPA environmental engineer, in a letter to members of Congress and the EPA inspector general, called that study "scientifically unsound," and questioned the bias of the panel, noting that five of the seven members had significant ties to the industry. "EPA's failure to regulate [fracking] appears to be improper under the Safe Water Drinking Act and may result in danger to public health and safety," he wrote.[96]

The following year, the Energy Policy Act of 2005[97] (P.L. 109-58)—by a 249–183 vote in the House and an 85–12 vote in the Senate—exempted the oil and natural gas industry from the Safe Water Drinking Act.[98] That exemption applied to the "construction of new well pads and the accompanying new roads and pipelines." The Natural Resources Defense Council noted that the EPA interpreted the exemption "as allowing unlimited discharges of sediment into the nation's streams, even where those discharges contribute to a violation of state water quality standards."[99] Vice-President Dick Cheney, whose promotion of Big Business and opposition to environmental policies is well-documented, had pushed for that exemption. His hand-picked "energy task force," composed primarily of industry representatives, had concluded that fracking was a safe procedure. Cheney had been CEO of Halliburton, one of the world's largest energy companies; the exemption became known derisively as the Halliburton Loophole. That legislation, says Al Gore, "put the whole industry in such a privileged position, it disadvantages the advocates of the public interest, which was the intention."[100]

Robert F. Kennedy Jr., an environmental lawyer who has been an active environmentalist more than three decades, was blunt in assessing the natural gas industry's lobbying and propaganda machine: "The industry's worst actors have successfully battled reasonable regulation, stifled public disclosure while bending compliant government regulators to engineer exceptions to existing environmental rules."[101]

Natural gas companies and their contractors are also exempt[102] from regulations of the National Environmental Policy Act,[103] and the nation's SuperFund law,[104] which requires companies that pollute the environment to take a fiscal responsibility. The natural gas industry, prior to establishing a well pad and fracking for gas, also doesn't worry about having to submit an Environmental Impact Statement, something required for most businesses in every state.

An editorial in the Oct. 19, 2011, issue of *Scientific American* pointed out the reality of the lack of adequate regulation, and the willingness of states to embrace the natural gas industry:

> "Fracking is already widespread in Wyoming, Colorado, Texas and Pennsylvania. All these states are flying blind. A long list of technical questions remains unanswered about the ways the practice could contaminate drinking water, the extent to which it already has, and what the industry could do to reduce the risks."[105]

Scientific American recommended that the federal government establish "common standards" and that "[S]tates should put the brakes on the drillers" until the EPA completes a preliminary study of the effects of the environmental effects of fracking.[106] The agency won't be looking at health impacts.

Bills introduced in the U.S. House (H.R. 2766)[107] by Reps. Diana DeGette (D-Colo.), John Salazar (D-Colo.), and Maurice Hinchey (D-N.Y.); and the U.S. Senate (S. 1215)[108] by Sens. Robert P. Casey Jr. (D-Pa.) and Chuck Schumer (D-N.Y.) in June 2009 would have given federal regulatory oversight under the Safe Water Drinking Act to hydraulic fracturing; however, the bills languished. New bills (H.R. 1084[109] and S. 587[110]), introduced in March 2011 in the 112th Congress, also died without a vote.

The extent of energy company influence is easily seen by a vote in the U.S. Senate in March 2012. President Barack Obama had called for an end to the $4 billion subsidy to oil companies.[111] The previous year, the three largest oil companies had earned about $80 billion in profits.[112] Over the past decade, the five largest oil companies had made $1 trillion in profits. "We have been subsidizing oil companies for a century. That's long enough," the President said.[113] But, the Senate disagreed. Forty-three Republicans and four Democrats blocked the elimination of subsidies for oil companies. Although the final vote was 51–47 to end the subsidies, a simple majority was not enough because the Republicans threatened a filibuster that would have required 60 votes to bring the bill to a vote. A Think Progress financial analysis revealed that the 47 senators who voted to continue subsidies received $23,582,500 in career contributions from the oil and gas industry. In contrast, the 51 senators who had voted to repeal the subsidies received $5,873,600.[114]

A month after the Senate voted to continue oil company subsidies, President Obama, who had long advocated for the development of natural gas as a temporary fuel to replace fossil fuel dependency, issued an executive order to coordinate oversight of the natural gas industry. The order affects several White House offices and cabinet departments, including the EPA. The natural gas industry and several Republican leaders had previously criticized the President for what they erroneously believed was his opposition to natural gas development, and had lobbied heavily for such an order to reduce what they saw as overlapping jurisdictions and over-regulation of the Industry.

"We are pleased to see this action today, which will help promote consistency between the Administration and policies that are put in place [and will improve] government communication and coordination that will help our members continue to safely deliver this foundation fuel to 177 million Americans every day,"[115] said Dave McCurdy, president of the American Gas Association. He also noted that the President "has been promoting responsible production and broader use of this domestic, abundant, affordable, clean and reliable energy source." Favorable statements also came from the Marcellus Shale Coalition, America's Natural Gas Alliance, American Petroleum Institute, Independent Petroleum Association of America, American Chemistry Council, and the Dow Chemical Co.[116] Even with an industry-favorable decision by the President, there was still sniping from the right-wing. "We don't need another working group, or any more bureaucracy," a spokesman for House Speaker John Boehner (R-Ohio) said in a prepared statement.[117]

President Obama later shunted aside his philosophy that natural gas could be a bridge fuel, a temporary solution to the energy crisis. In his 2013 State of the Union address, while calling for better research and technology, he declared:

"[T]he natural gas boom has led to cleaner power and greater energy independence. We need to encourage that. And that's why my administration will keep cutting red tape and speeding up new oil and gas permits."[118]

CHAPTER 2
Following the Money

The oil and natural gas industries have a history of effective lobbying at the state and national levels. The Natural Gas Alliance has four former congressmen as lobbyists, according to the Center for Responsive Politics (CRP).[119] Between 1990 and October 2012, the oil and gas industry (PACs, individuals, and outside soft money) contributed about $238.7 million to candidates for elected positions; about three-fourths of the contributions went to Republican candidates, according to CRP data based upon reports of the Federal Elections Commission (FEC).[120]

For the 2008 election cycle (Jan. 1, 2007–Dec. 31, 2008), contributions were about $37.9 million, but dropped in the off-year election cycle of 2009–2010 to $32.1 million. Total contributions for the 2011–2012 election cycle were $63.6 million, with almost 90 percent of it going to Republicans.[121]

Contributions in the 2011–2012 cycle by PACs and individuals associated with the oil and gas industry to Mitt Romney, the Republican presidential nominee, as of the FEC report filed Nov. 12, 2012, the week after the election, was $4,763,934.[122] During one of the televised debates, Romney stated the EPA was "a tool in the hands of the President to crush the private enterprise system," and suggested that regulation of fracking should be left solely to the individual states.[123] His faulty reasoning was "the EPA and those extreme voices in the environmental community and in the President's own party are just frustrated beyond belief that the states have the regulatory authority over fracking."[124]

Other contributions to Republican candidates went to Rick Perry ($969,824), Ron Paul ($178,121), Newt Gingrich ($139,410), Herman Cain ($122,147), Rick Santorum ($91,925), Tim Pawlenty ($52,050), Michele Bachman ($47,140), and Jon Huntsman ($32,000). Gary Johnson, the Libertarian candidate, received $10,700; Dr. Jill Stein, the Green Party candidate, received $750 from the oil and gas industry.[125]

President Barack Obama received $710,277. During the 2008 presidential campaign, Obama had received $952,900 from the oil and gas industry.[126] His opponent, Sen. John McCain, accepted $2,599,892.[127]

The natural gas industry contributed $3.7 million to current members of the House Energy and Commerce Committee, according to CRP.

The top recipients of oil and gas contributions during the 2012 election for House or Senate, according to the CRP, were Rep. Richard "Rick" Berg (R-N.D.), a member of the Ways and Means Committee and recipient of the Petroleum Council's Legislator of the Year award in 2009 ($433,949); Rep. John Boehner (R-Ohio), speaker of the House ($407,699); Senate Majority Leader Mitch McConnell (R-Ky.) ($395,950); Sen. Orin Hatch (R-Utah), a member of the Subcommittee on Energy, Natural Resources, and Infrastructure ($370,650); Rep. Denny Rehberg (R-Mont.) ($373,469); Sen. John Barrasso (R-Wyo.), a member of the Energy and Natural Resources Committee ($339,716); Rep. Mike Pompeo (R-Kans.), a member of the Energy and Commerce committee ($286,300); Rep. Eric Cantor (R-Va.), House majority leader ($245,250); Rep. Kevin McCarthy (R-Calif.), majority whip ($235,700); Rep. Dave Camp (R-Mich.), chair of the Ways and Means committee ($222,250); and Sen. Dean Heller (R-Nev.), a member of the Energy and Natural Resources Committee ($208,500).[128]

All 20 of the top recipients of oil and gas donations in both 2010 and 2012 election cycles were Republicans.

Rep. Lou Barletta (R-Pa.) was not among the top 20 recipients, but he is on the list of the top third of House members to receive campaign donations from individuals and PACs associated with the gas industry. As a candidate running against Paul Kanjorski, a 13-term member of Congress, Barletta received $31,725 in donations in the 2009–2010 election cycle.[129] For the 2011–2012 cycle, now a freshman representative, Barletta received $29,450.[130] Barletta is a member of the Congressional Marcellus Shale Coalition—and, according to reporting by Rick Dandes of the Sunbury (Pa.) *Daily Item*, "owns stock in [eight] gas industry companies, including several that are actively drilling in Pennsylvania and beyond."[131] Depending upon market fluctuation, value of the stock is more than $75,000. However, as reported by the *Daily Item*:

> "[T]here are no rules in place to bar him from buying stock. The practice is both legal and permitted under the ethics rules that Congress has written for itself, which allow law-makers to take actions that benefit themselves or their families except when they are the lone beneficiaries.
>
> "The financial disclosure system Congress has implemented also does not require the lawmakers to identify potential conflicts when they take official actions that intersect or overlap with their investments."[132]

Barletta's communications director told the *Daily Item* that Barletta "has an incredibly diversified portfolio" that includes non-energy stocks.[133]

The Republicans' determination to support the natural gas industry was apparent in how they treated President Obama's nomination of Rebecca Wodder, a biologist and environmental scientist, to be assistant secretary of the Department of Interior. Wodder had previously stated that fracking "has a nasty track record of creating a toxic chemical soup that pollutes groundwater and streams." Shortly after her nomination in June 2011, Sen. James Inhofe (R-Okla.), ranking member of the Senate Committee on Environment and Public Works, and one of the leading recipients of energy industry campaign funds,[134] declared he had:

> "serious concerns about her nomination: she is the CEO of a far-left environmental organization [American Rivers] and was a staunch supporter of the Clean Water Restoration Act, a bill that would have given the federal government authority over practically every body of water in the country, no matter how small. She is also an active proponent of federal regulation of hydraulic fracturing—a practice that is efficiently and effectively regulated by States."[135]

Several weeks later, before a Senate Energy and Natural Resources hearing, Wodder refused to retract her statement about fracking. This led Inhofe to declare Wodder was "beholden to an extremist environmental agenda,"[136] and lead a campaign against her nomination. Although the Senate Environmental and

Public Works Committee, on a strict party vote, later approved her nomination, there were not enough votes for her recommendation by the Energy and Natural Resources Committee.[137] In December 2011, the nomination was sent back to the White House; a month later, the Obama administration reluctantly did not re-nominate her.

In triumph, Inhofe boasted, "I am pleased that through our rigorous oversight, we succeeded in preventing Rebecca Wodder, another member of President Obama's job-destroying 'green team,' from assuming an influential position within the administration."[138] An Obama official, who asked not to be identified, had a different opinion. He said, "The fact that Rebecca Wodder, a highly qualified nominee, could not get confirmed is another reflection of how some Republicans have ground the Congress to a halt. If the nomination process wasn't so politically supercharged, Rebecca would have been confirmed months ago."[139] Wodder was then appointed senior advisor to the Secretary of the Interior, a position that did not need Senate confirmation.

In contrast to the Republicans, and some Democrats, taking energy industry money and not looking at the effects of drilling, the 1.4 million member Sierra Club, since August 2010, has refused to accept any donations from the natural gas industry. The Sierra Club, which actively opposes the development of coal as an energy source, had received $27 million between 2007 and 2010 from Chesapeake Energy, but may have deliberately withheld that information from its members.[140] By 2010, "our view of natural gas [and fracking] had changed [and we] stopped the funding relationship between the Club and the gas industry, and all fossil fuel companies or executives," said Michael Brune, who became Sierra's executive director in March 2010, succeeding Carl Pope who had accepted the Chesapeake donations for the Sierra Club.[141] However, the Sierra Club, although it no longer believes natural gas is a "bridge fuel," still calls for a moratorium rather than an outright ban.[142]

Political influence through lobbying increased significantly because of the Supreme Court's 5–4 decision in *Citizens United v. Federal Election Commission*.[143] That decision, handed down in January 2010, gave corporations the same First Amendment rights as individuals. Thus, corporations could create significant campaign assistance for political candidates, including producing campaign ads for print and electronic media. The decision was supported by the U.S. Chamber of Commerce, the National Rifle Association, and most conservative organizations. In response, Barack Obama said that the *Citizens United* decision "gives the special interests and their lobbyists even more power in Washington—while undermining the influence of average Americans who make small contributions to support their preferred candidates."[144] Later that month, in his State of the Union address, the President was even more forceful, arguing, "The Supreme Court reversed a century of law to open the floodgates for special interests—including foreign corporations—to spend without limit in our elections. Well I don't think American elections should be bankrolled by America's most powerful interests, or worse, by foreign entities."[145]

Chevron, the third largest U.S. corporation, took advantage of the *Citizens* decision to contribute $2.5 million in October 2012 to the Congressional Leadership Fund,[146] which supports conservative politicians; almost all of those they support are in favor of expanded oil and natural gas drilling, and have no objection to the use of fracking. Chevron also acknowledged spending $9.5 million solely in lobbying costs for members of Congress in 2011.[147]

FOLLOWING THE MONEY INTO PENNSYLVANIA

Mixed into Pennsylvania's energy production is not only a symbiotic relationship of business and government, but also a history of corruption and influence-peddling in all of its 67 counties.

The Keystone State, with no limits on political contributions, is also the target of energy company generosity to politicians. The natural gas industry, through PACs and individuals associated with the natural gas industry, contributed about $8 million to Pennsylvania candidates and their PACs between 2000 and August 2012, including $860,825 to the Republican party and $129,100 to the Democratic party, according to Common Cause.[148] Total lobbying expenditures, in addition to direct financial contributions between 2007 and 2012 were $15.7 million.[149]

Of the top 15 members of Congress from Pennsylvania who

accepted campaign funds from PACs and individuals affiliated with energy companies, 11 were Republicans, according to Common Cause. Rep. Tim Murphy (R-Pa.), with $275,499, was tenth among all Congressional recipients. Murphy is vice-chair of the Subcommittee on Environment and Economy and a member of the House Energy and Commerce Committee. Sen. Pat Toomey (R-Pa.), with $160,750, ranked 28th nationally of all recipients.

At the state level, the natural gas industry contributed about $1.8 million to Tom Corbett's political campaigns between 2000 and April 2012,[150] about $1.1 million of that for his campaign for governor.[151] An additional $6 million may have been funneled through the Republican Governor's Association.[152] The first influx of energy company money came in Corbett's first run for attorney general. Will Bunch, columnist for the *Philadelphia Daily News*, revealed the long connection between Corbett and the industry:

> "The $450,000 in campaign checks that energy mogul Aubrey McClendon [CEO of Chesapeake Energy] wrote that fall [2004] helped elect a man he said he'd never even met—a relatively obscure GOP candidate for Pennsylvania attorney general, Tom Corbett. . . .
>
> "That investment arguably changed not just the history but also the political direction of the state. The influx of cash helped Corbett narrowly win the closest attorney general's race in Pennsylvania history and propelled him toward the governor's mansion. . . .
>
> "Did Oklahoma gas driller McClendon see the coming boom in drilling in the gas-rich, Pennsylvania-centered formation known as the Marcellus Shale back in 2004? And did he see his massive campaign contributions—filtered through an obscure GOP committee—as a shrewd down payment on future political access and influence?
>
> "Or was it merely a case of what McClendon and Chesapeake officials have maintained all along—that the energy millionaire was simply writing so many checks for conservative causes that year, including $250,000 for the notorious John Kerry-bashing Swift Boat Veterans for the Truth, that he wasn't even aware that his cash was going to the state's future top prosecutor?"[153]

Chesapeake Energy, which owns drilling rights to about 15 million acres in the U.S., and about 1.5 million acres in the Marcellus Shale,[154] with 1,935 permits[155] became the state's leading producer of natural gas,[156] and also a company with a spotty history of environmental concerns.[157] By the beginning of 2013, Chesapeake had the most penalties (485) of any company drilling in the state.[158]

Tom Corbett's first major political appointment after his election as governor in November 2010 was to name C. Alan Walker, an energy company executive, to head the Department of Community and Economic Development. The *Pennsylvania Progressive* identified Walker as "an ardent anti-environmentalist and someone who hates regulation of his industry."*[159]* A *ProPublica* investigation revealed that Walker had given $184,000 to Corbett's political campaign.[160]

In October 2010, Gov. Ed Rendell, by executive order, had imposed a moratorium on drilling on public lands.[161] However, shortly after taking office, Corbett repealed environmental assessments of gas wells in state parks, but did not lift the moratorium. Sixty-one of the 120 state parks sit above the Marcellus Shale.[162] Further, Corbett decided that about half of the 1.5 million acres of state forest lands over the Marcellus Shale could be leased for mineral rights.[163] The result could be as many as 2,200 well pads on almost 90 percent of all public lands, according to Nature Conservancy of Pennsylvania,[164] which argued that deforestation from drilling would alter and destroy the ecological balance.

In October 2012, John Norbeck, the widely-respected head of the state's park system, abruptly resigned. The *Harrisburg Patriot-News* reported:

> "Environmentalists and advocates for the state parks fear Norbeck's abrupt departure could be a sign that the Corbett administration is preparing to open the park gates to drilling rigs to tap natural gas under the Commonwealth's popular park system; Norbeck is known to be an opponent of drilling in the state parks."[165]

Corbett's public announcements in March 2011, two months after his inauguration, had established the direction for gas drilling in Pennsylvania. In his first budget address, Corbett declared he wanted to "make Pennsylvania the hub of this [drilling] boom. Just as the oil companies decided to headquarter in one of a dozen states with oil, let's make Pennsylvania the Texas of the natural gas boom. I'm determined that Pennsylvania not lose this moment."[166] Still enthusiastic about fracking Pennsylvania, Corbett told an industry-sponsored conference in Philadelphia in September 2012, "I am convinced that we are at the beginning of a new industrial revolution and you are at the tip of the spear."[167]

Within his first budget bill, Corbett had authorized Walker to "expedite any permit or action pending in any agency where the creation of jobs may be impacted."[168] This unprecedented reach apparently applied to all energy industries. That same month, Corbett created the Marcellus Shale Advisory Commission, which he loaded with persons from business and industry. Not one member was from the health professions; of the seven state agencies represented, not one member was from the Department of Health. Lt. Gov. Jim Cawley, who chaired the Commission, boasted, "The Marcellus [Shale] is revitalizing our main streets in downtowns."[169] Like Corbett, he didn't say anything about health and environmental effects.

The same month Corbett delivered his first budget address, Michael Krancer, the new DEP secretary, a political appointee whose wife and father donated a combined $306,500 to Corbett's 2010 political campaign,[170] took personal control over his department's issuance of any violations. By Krancer's decree, every inspector could no longer cite any well owner in the Marcellus Shale development without first getting the approval of Krancer and his executive deputy secretary.

"It's an extraordinary directive [that] represents a break from how business has been done" and politicizes the process, John Hanger told *ProPublica*.[171] Hanger, DEP secretary in the Ed Rendell administration, said the new rules "will cause the public to lose confidence entirely in the inspection process."[172] The new policy was the equivalent of every trooper having to get permission from the state police commissioner before issuing a traffic citation, Hanger said.[173] Because the new policy is so unusual and broad "it's impossible for something like this to be issued without the direction and knowledge of the governor's office," said Hanger. Corbett denied he was responsible for the decision. Five weeks after Krancer's decision was leaked to the media and following a strong negative response from the public, environmental groups and the media, the DEP rescinded the policy—which Krancer now claimed was only a three-month "pilot program."[174]

On Tuesday, Feb. 14, 2012, Tom Corbett, surrounded by Republicans, gave the natural gas industry a Valentine's Day gift, and proudly signed Act 13 of 2012, an amendment to Title 58 (Oil and Gas Act; 58 P.S. §601.101 *et seq.*)[175] of the Pennsylvania Consolidated Statutes. Pennsylvania's new law that regulates and gives favorable treatment to the natural gas industry was initiated and passed by the Republican-controlled General Assembly. The House had voted 101–90 for passage;[176] the Senate voted, 29–20.[177] Both votes were mostly along party lines. Environmental and conservation groups spent under $50,000 to lobby against the proposed act; the natural gas industry spent about $1.3 million for it, according to data compiled by Dory Hippauf.[178]

"Thanks to this legislation," said Corbett upon signing the bill, "this natural resource will safely and fairly fuel our generating plants and heat our homes while creating jobs and powering our state's economic engine for generations to come."[179]

Rep. State Sen. Charles T. McIlhinney (R-Doylestown) was effusive in his praise:

> "As we worked to craft a compromise, protecting the environment and preserving local zoning control were two of my primary concerns. The final legislation accomplished these goals, earning the support of Pennsylvania Association of Township Supervisors, other local government organizations and the state's Growing Greener Coalition. While not perfect, it is a balanced and thoughtful approach to protecting our environment and regulating an industry that is here to stay in Pennsylvania."[180]

McIlhinney may have believed what he said, but he was wrong on almost every point.

Act 13 is generally believed to be "payback" by Corbett and the Republican legislators for campaign contributions. "The industry has largely had its way in Pennsylvania and has spent millions to put their friends in the state legislature and the Governor's mansion," said James Browning, Common Cause regional director of state operations.[181] The focus for the oil and gas industry, says Browning, is on protecting these investments and maintaining access to key elected officials."

Rep. Brian L. Ellis (R-Butler County), sponsor of HB 1950, the base for Act 13, received $23,300 from PACs and individuals associated with the oil and gas industries. Sen. Joseph B. Scarnati (R- Warren, Pa.), the senate president *pro-tempore* who sponsored the companion Senate bill (SB 1100), received $359,145.72 as of April 2012, according to Marcellus Money.[182] Rep. Dave Reed, chair of the majority policy committee, received $137,532.33; Rep. Mike Turzai, majority floor leader, received $98,600; Sen. Don White (R-Indiana), a member of the Environmental Resources and Energy committee, received $94,150; Sen. Jake Corman (R-Centre County), chair of the Appropriations Committee, received $91,290. Of the 20 Pennsylvania legislators who received the most money from the industry between 2001 and 2012, 16 are Republicans, according to Common Cause.[183]

Rep. H. William DeWeese (D-Waynesburg, Pa.), received $53,300, the most of the four Democrats at the time of the vote for what became Act 13. DeWeese, first elected in 1976, had been Speaker of the House and Democratic leader. In April 2012, DeWeese was sentenced to 30 to 60 months in prison for theft, conspiracy, and conflict of interest, all related to the use of legislative staff and public resources for campaign work.[184] He was also ordered to pay $25,000 in fines and about $117,000 in restitution. DeWeese maintained that the prosecution, begun while Tom Corbett was attorney general, was political.[185] DeWeese had charged that Corbett's prosecution of state legislators, at first primarily Democrats, was to set a base for Corbett's run for governor. However, DeWeese also charged that Corbett specifically targeted him because he opposed Corbett's belief that natural gas companies should not pay an extraction fee.[186] DeWeese, whose district includes about 2,000 wells, increased his criticism after Corbett became governor.

Gov. Ed Rendell had wanted a 5 percent severance tax on the value of gas produced, plus a fee of 4.7 cents for every 1,000 cubic feet in order to create additional revenue for the general fund. However, he couldn't get the support of the Legislature. Part of the problem was that opponents of the tax cited a Penn State study that claimed a 30 percent decline in drilling if the fees were assessed, while also touting the economic benefits of drilling in the Marcellus Shale. What wasn't widely known is that the lead author of the study, Dr. Timothy Considine, "had a history of producing industry-friendly research on economic and energy issues," according to reporting by Jim Efsathioi Jr. of *Bloomberg News*.[187] The Penn State study was sponsored by a $100,000 grant from the Marcellus Shale Coalition, a coalition of about 300 energy companies which says it provides "in-depth information to policymakers, regulators, media, and other public stakeholders on the positive impacts responsible natural gas production is having on families, businesses, and communities across the region."[188] Dr. William Easterling, dean of Penn State's College of Earth and Mineral Sciences, said the study may have "crossed the line between policy analysis and policy advocacy."[189]

If Pennsylvania was to impose a severance tax, the natural gas industry had a plan. According to Karen Feridun of Berks Gas Truth:

> "They were already saying that they were too poor to set up operations *and* pay a severance tax, so they planned to lobby for getting a 3–5-year window to set up before starting to pay the tax. Drilling is like wringing a sponge. You get the most liquid out of a sponge the first few times. You get more out in subsequent wrings, but nothing like what you got the first couple of times. The drillers planned to drill like crazy in those first 3–5 years, knowing that they'd be paying next to nothing when the tax finally took effect."

The natural gas industry didn't need to worry. Tom Corbett, pushing hard for the gas industry, had originally wanted no tax or impact fees placed upon natural gas drilling;[190] however, as public discontent increased, he suggested a 1 percent tax, which was in the original House bill. In contrast, other states that allow natural gas fracking have tax rates as high as 7.5 percent of market value (Texas) and 25–50 percent of net income (Alaska).[191] The Pennsylvania rate can vary, based upon the price of natural gas and inflation, but will still be among the five lowest of the 32 states that allow natural gas drilling. Over the lifetime of a well, Pennsylvania will collect about $190,000–$350,000, while West Virginia will collect about $993,700, Texas will collect about $878,500, and Arkansas will collect about $555,700, according to Pennsylvania Budget and Policy Center (PBPC) data and analyses.[192]

State Rep. Mark Cohen, a Democrat from Philadelphia, like most of the Democrats in the General Assembly, opposed the proposed legislation, which he says, "produces far too little revenue for local communities, gives the local communities local taxing power which most of them do not want, because it pits one community against the other, and gives no revenue at all to other areas of the state." State Sen. Daylin Leach, a Democrat from suburban Philadelphia, agrees. "At a time when we are closing our schools and eliminating vital human services, to leave billions on the table as a gift to industry that is already going to be making billions is obscene."

Each well is expected to generate about $16 million during its lifetime,[193] which can be as few as ten years, according to the PBPC. The effective tax and impact fee is about 2 percent, with a minimum royalty to landowners of 12.5 percent.

"Pennsylvania politicians sold gas companies the right to pollute Pennsylvania's land, air, and water for bargain basement prices," said Josh McNeil, executive director of Conservation Voters of Pennsylvania, who noted, "For their $23 million political investment, gas companies avoided hundreds of millions in taxes that could have paid for thousands of teachers, roads and desperately needed environmental protections."[194]

At the time Corbett signed Act 13, the state had a $4 billion deficit,[195] which could have been significantly reduced had Pennsylvania imposed fees and taxes in line with other states. However, even *if* there was an adequate severance tax, should it have been allocated to the general fund or used primarily to repair the damage caused by the natural gas industry?

In September 2012, the Pennsylvania Public Utility Commission announced it billed energy companies $206 million for impact fees from 4,453 wells for 2011.[196] The energy industry is believed to have earned about $3.5 billion in that year from

40

wells in Pennsylvania, according to a previous AP analysis.[197] The impact fee was $50,000 per non-conventional well and $10,000 for each conventional well. That fee varies depending upon the price of natural gas. Chesapeake Energy paid $30.8 million for 624 wells; Talisman Energy paid $26.4 million on 540 wells; Range Resources paid $23.7 million on 475 wells. The PUC reported that the state was taking $25 million; 40 percent of the remainder was to go to 37 counties and 1,500 municipalities that had wells, and 60 percent split among state agencies that dealt with drilling. Corbett, of course, had not wanted any fees or taxes imposed upon the natural gas Industry; Gov. Rendell's proposal would have brought in as much as $500 million.

In contrast to giving favored status to the natural gas industry and proposing significant tax cuts for business, in his first budget message Tom Corbett had pushed for draconian cuts in education, health care for the poor, child care, and services to the disabled. The 2012–2013 budget, readily passed by the Republican-controlled legislature, reflected even deeper cuts to health and human services; environmental protection was cut $2 million from the previous year; the Department of Conservation and Natural Resources also took a hit.

In his first two annual budgets, Tom Corbett had cut funding to the Departments of Environmental Protection and Conservation and Natural Resources. For the 2012–2013 fiscal year, Corbett cut DEP funding about $10.8 million, a 7.8 percent decrease, reducing the state appropriation to about $124 million.[198] However, the DEP also received a $40 million cut in its $268 federal appropriation from the previous year, leaving it with about $51 million less to work with.

In a commentary published in the Harrisburg (Pa.) *Patriot-News*, State Rep. Greg Vitali (D-Delaware County), pointed out, "Nonunion [DEP] staff has not received a salary increase in four years. Noncompetitive salaries combined with increaseing workloads due to these staffing cuts have made it difficult for the DEP to attract and retain quality people."[199] Vitali says he was told by a former senior staff member of the DEP, "We are hemorrhaging jobs to the oil and gas industry."[200]

Michael Krancer claimed the decreased funding would not

affect the department's well inspections. He told the House Appropriations Committee in March 2012 that the inspections program "has been a function of the permit fees," and there was increased staff to deal with the increased number of wells.[201] Eighty inspectors conducted 5,000 field inspections in 2011. What Krancer did not say is that each inspector had increased workloads, and there was a paperwork backlog in processing permit fees. (Nationally, between 2010 and 2011, the number of drilling rigs increased by 22 percent, but the number of inspecttions fell by 12 percent, according to an analysis by the *New York Times* of 50,000 inspection reports.[202])

Instead of adding staff to take care of that backlog, Corbett issued an industry-friendly executive order in July 2012. That order directed the DEP to "establish performance standards for staff engaged in permit reviews and consider compliance with the review deadlines a factor in any job performance evaluations."[203] That requirement, said George Jugovic Jr., former director of DEP's Southwest Region and current president of PennFuture, is "going to put the public health and safety at risk." Jugovic told the Associated Press, "I think the message is clear. Issuing the permit has a higher priority than doing a fair and thorough job of insuring that the application complies with the law." The order, said Jugovic, "does not recognize any of the complexities of what the agency is required to do [except to] beat down an already demoralized staff."[204] Statements made in a closed-door legal proceeding the previous year revealed that because of political and administrative pressures, DEP staff spent few as 35 minutes per application, with supervisors spending as few as two minutes to review each application.[205] "Such a cursory review leaves little time to consider and include necessary permit provisions or technical requirements to protect public health and the environment," Lisa Sumi wrote in *Breaking All the Rules* (September 2012), a 124-page summary of extensive field research conducted by Earthworks.[206] Sumi, an environmental consultant and author of several major research studies on energy, noted, "In Pennsylvania, citizens have conducted research and file reviews that have exposed deficiencies in permits [but] citizens do not have the resources to review all permits, nor should they be doing the work that agencies are charged to do."[207]

Three months later, the DEP established a policy to delay notifying the public of water contamination. The DEP had previously issued "notices of contamination" to the public as soon as DEP scientists had made the determination. The new policy restricts notification until after DEP executives are notified and approve the release of information. "This change in procedure is unnecessary and a dangerous policy," said Karen Feridun of Berks Gas Truth. "With this secretive change in policy," without giving the public time to comment, said Thomas Au of the Sierra Club, "the DEP has violated fundamental democratic values of transparency and public participation."[208]

An additional blow to scientific integrity and the protection of the environment is HB 1659[209] that is meandering its way through the legislative process. Introduced by Rep. Jeffrey P. Pyle (R-Armstrong and Indiana counties), the bill would symbolically put foxes in charge of the henhouse. The bill calls for automatic approval of permits if the DEP can't approve them within a limited time frame. The bill also requires that the DEP create a system to allow individuals not employed by the DEP to review permits. If passed into law, the outside review "will encourage the Gas Industry to go Permit Review Shopping to find the 'reviewer' with the fastest and cheapest rubber stamp," Dory Hippauf wrote in her blog, *Fracktoids*.[210]

"U.S. regulators at the state level have been captured by the industry," observed Robert F. Kenedy Jr.[211]

Psychologist Diane Siegmund is frustrated by what she sees not only as state government's acceptance of fracking but also of numerous local governments in the Marcellus Shale region not acting on behalf of the preservation of health and the environment. When she went to the Bradford County commissioners with stacks of research about problems with fracking, "all they did was to thank me and claim it's not their problem." She says residents are beginning to believe that local governments are operating in collusion with the energy companies. The attitude of Big Government being hostile to the people is also a reason why many who have sustained damage do not speak up. "Perhaps it is against their culture to do so," says Siegmund, "or maybe it is just because they see no way to get it made 'right.' "

She says there "is a lot of fear" among the residents, those whose lives are being uprooted, those whose health is being compromised, and those whose economic benefits may be compromised if fracking operations are reduced. "As long as the powers can keep the people isolated and fragmented," says Siegmund, "the momentum for change can never be gained."

Lisa Sumi, an environmental scientist who had been research director of Earthworks: the Oil and Gas Accountability Project (OGAP), and the person who coined the term "fracking" in 2004, believes:

> "This betrayal of the public interest also severely weakens state claims that they can protect the public from the impacts of the shale boom. A rule–even an improved rule–on the books means little if an oil or gas company knows that it can be ignored with little or no consequence."[212]

"When state agencies say they will 'regulate' or 'monitor' hydraulic fracturing, we should not accept this as a guarantee of any kind," says Eileen Fay, an animal rights/environmental writer. Fay argues that because of legislative corruption, it is a responsibility of citizens to protect their own health and environment by "putting pressure on our legislators."

FRACKING EDUCATION

The 2012–2013 education budget remained the same as the previous year, having already suffered more than $1 billion in cuts in the 2011–2012 budget. However, Corbett had a strange idea how the 14 state-owned universities could restore some of their budget he had proposed be cut by half.[213] In April 2011, Corbett had suggested that the State System of Higher Education (SSHE) could allow natural gas drilling on the campuses that sit on top of the Marcellus Shale. Several months later, the state Senate passed a bill sponsored by Donald C. White (R-Indiana) that authorized state officials to lease mineral rights beneath state land to gas, oil, and coal companies. White was the recipient of $94,150 in donations from PACs and individuals specifically associated with the natural gas industry, according to Pennsylvania Common Cause.[214] The Senate passed the Indigenous Mineral Resource Development Act (SB 367[215]), 46–3; the House passed the bill, 136–62. Although only six of the 14 universities are in the Marcellus Shale, the bill permits drilling for oil, coal, coal-bed methane, or limestone on all university campuses. Corbett signed the bill in October 2012.

In an attempt to placate the SSHE, the new act allows the university where the gas is extracted to retain one-half of all royalties; 35 percent would go to the other state universities; 15 percent would be used for tuition assistance at the 14 state universities.

California University of Pennsylvania, a SSHE institution about 35 miles south of Pittsburgh, already has ceded mineral rights to the natural gas industry. In a secret negotiation revealed by the *Pittsburgh Post-Gazette*, the Student Association, which owns recreational and dormitory space at the university, signed a lease in January 2011 with Antero Resources Appalachian for subsurface drilling rights on 67 acres.[216] The lease included a confidentiality clause. The *Post-Gazette* reported in November 2011 that the Association initially refused to say if it had such a lease, "even though its offices are on campus and its executive director is a CalU employee."[217] That executive director was the university's dean of students and acting vice-president for student affairs. The university's president told the *Post-Gazette* he didn't know about the lease. However, he had previously suggested to Antero Resources that since the university wasn't authorized to enter into any lease, it might wish to contact the student association.

The SSHE chancellor's office supports drilling on or near state universities, and has been actively working with the natural gas industry to create programs that prepare students to enter the shale gas industry. The Marcellus Institute at Mansfield University, located in Tioga County, which has the second highest number of wells in the state, is "an academic/shale gas partnership," designed to educate the people about the issues of natural gas production. The university even held a three-day summer camp for high school students in July 2012 to allow students to "Learn about the development of shale gas resources in our region and the career and educational opportunities available to you after high school!"[218]

The university's associate in applied sciences (A.A.S.) degree in natural gas production and services, begun in Fall semester 2012, has five separate tracks—Permitting and Inspection,

Mudlogging/Geologic Technician, Environmental Technician, GIS Technician, and Safety Management. The degree was fast-tracked, submitted and approved in less than six months, rather than the 12–18 months normally required for approval. "The industry is here and now, and we needed to take advantage of that," says Lindsey Sikorski, acting director for the program. She says the program began when the university and SSHE took notice that 40 of its graduates, most in geology and geography, were working in the industry.

The university "will take as many students as we can," says Sikorski, although only one new faculty position was approved. The SSHE administration encouraged larger class sizes and fewer permanent professors. The program, Sikorski says, "is not one of advocacy for the industry, and all sides will be considered." The program has not received any grants from the industry; Sikorski says she "doesn't want there to be any conflicts of interest" that would "compromise the integrity of the program." However, the reality is that energy companies and their lobbying groups may eventually fill a hole created by Tom Corbett slashing higher education funding and Chancellor John C. Cavanaugh refusing to protect academic integrity in the state-owned universities.

At Indiana University of Pennsylvania, the energy resources track in the Department of Geoscience "will prepare students for direct entry into the energy industry with a focus on the discovery and development of energy resources and geophysical exploration techniques." The university added two additional faculty, and anticipates about 20 geology students will select that track option.

The Shale Technology and Education Center (ShaleTEC) program at the Pennsylvania College of Technology (Williamsport, Pa.), a branch of Penn State, was established, according to its website, "to serve as the central resource for workforce development and education needs of the community and the oil and natural gas industry." All courses at ShaleTec "may be customized to meet your company's needs." Many courses are half-day or full-day lectures/seminars. The college also offers several degrees in fields allied to the natural gas industry.

The Community College of Philadelphia (CCP)—assisted by the state's Department of Labor and Industry and the Marcellus Shale Coalition—created the Energy Training Center in November 2012 to offer certificate and academic programs for workers either already employed by or intending to enter jobs that provide services to Marcellus Shale companies. The Coalition gave CCP an initial $15,000 grant for scholarships.

In a news release loaded with pro-Corbett and pro-industry appeal, college president Stephen M. Curtis announced, "The goal is to support the supply chain now serving energy companies and offer specialized career training that connects residents to the high-pay, high-demand career paths."[219]

Creation of the Center "is short-sighted and foolhardy [since] we now know that shale gas drilling actually accelerates climate change," warned Margaret Stephens, associate professor of environmental conservation and geography.

Several faculty questioned the accelerated pace the program was initiated and the failure to inform most staff and faculty more than a day before the public announcement. Dr. Miles Grosbard, head of the Department of Architecture, Design and Construction, charged:

> "Normal college procedures for instituting new academic curricula were completely sidestepped. There is no information available about the proposed unit's mission, student audience, administrative structure, budget, facilities or educational objectives, apparently because none exists. Moreover, $15,000 is an impossibly tiny endowment to even begin a training center."[220]

John Braxton, assistant professor of biology and an ecologist, said CCP "must not be used as a PR puppet for shale gas fracking companies," accurately noting that the fracking industry "got a free publicity ride" by the administration's hasty decisions.

Within two weeks of CCP's announcement, the faculty union (AFT Local 2026), which represents the college's 1,050 faculty and 200 staff, condemned the decision to establish the Center "without the consideration or approval of the faculty, and with total disregard for established College procedures for instituting new academic curricula." In a unanimous vote by the Representative Council, the faculty declared, "the natural gas drilling . . . industry and peripheral and related industries present unacceptable dangers and risks to public health, worker safety, the natural environment, and quality of life." The faculty called for the college administration to "sever all ties to the Marcellus Shale Coalition, halt the implementation of any workforce training efforts related to the shale-gas fracking industry, and insist that any energy training programs be disassociated from the fracking industry and decided through the normal College governance process." In contrast to pushing training and education to benefit the fracking industry, the faculty urged the college "to expand its initiatives and offerings in clean, green energy, and environmental career fields."[221]

Several other colleges throughout the country, realizing they could make quick money from leases, have allowed the gas industry onto their campuses. In West Virginia, both Bethany College and West Liberty University signed leases, claiming the money from royalties would help improve programs and provide for new buildings.[222] The University of Texas at Arlington has about two dozen wells on its campus.

CORPORATE WELFARE FOR THE GAS DRILLERS

Continuing to promote fracking while cutting necessary social programs, draining state fiscal resources, and planning to pockmark college campuses to benefit the natural gas industry, Tom Corbett extended benefits to a foreign corporation, which was thinking about building an ethane cracker plant about 30 miles northwest of Pittsburgh. A cracker plant takes natural gas and breaks it up to create ethylene, primarily used in plastics. Royal Dutch Shell, which owns or leases about 900,000 acres in the Marcellus Shale basin,[223] considered placing the plant beside the Ohio River in Pennsylvania, Ohio, or West Virginia. All three states were interested, but Pennsylvania held out the most lucrative corporate welfare check. The Pennsylvania legislature handed over a 15 year exemption from local and state taxes, apparently without consulting local officials in Beaver County's Potter and Center townships.[224] Tom Corbett then approved a $1.65 billion tax credit over 25 years, tweeting, "A crackerplant would create up to 20,000 permanent jobs in Southwest PA."[225] However, the reality is considerably lower. Shell stated it planned to hire only 400 to 600 persons; because of the location, many new employees would probably be Ohio and West Virginia residents. Even if all possible indirect jobs—including more low-wage clerks at local fast food restaurants—were added, the most would be about 6,000–7,000 employees.

Pennsylvania may have been able to attract the plant without giving up so much corporate welfare. A Shell news release stated the company "looked at various factors to select the preferred site, including good access to liquids rich natural gas resources, water, road and rail transportation infra-structure, power grids, economics, and sufficient acreage to accommodate facilities for a world scale petrochemical complex and potential future expansions."[226] Even then, Shell said it could be "several years" before construction would begin. At the proposed location, the Horsehead Corp., which signed an agreement with Shell, has until April 30, 2014, before Shell could begin construction.

Dory Hippauf's "Connecting the Dots" series explains why Corbett may have been so generous with extending tax credits and subsidies. Between 2004 and 2011, Shell had donated about $358,000 to Corbett's campaigns for attorney general and governor, and persons with connections to Shell served on the state's Marcellus Shale Commission.[227]

EXEMPTING THE AFFLUENT SUBURBS

Pennsylvania politics continued to play out in how the Republicans separated one of the wealthiest and more high tech/industrial areas of the state from the rural areas. Less than a week before the 2011–2012 fiscal year budget was scheduled to expire, the majority party slipped an amendment into the 2012–2013 proposed budget, (SB1263[228]), to ban drilling in a portion of southeastern Pennsylvania for up to six years in the South Newark Basin, a rift basin that includes parts of Bucks, Montgomery, Berks, Chester, and Lehigh counties. That basin could provide at least 360 billion cubic feet of natural gas.[229]

Less than two days before the new budget would be voted upon, anti-fracking activist Iris Marie Bloom sent an e-mail blast; AP later transmitted a short article to inform the public of the new amendment that was passed and signed into law by Gov. Corbett, June 30, 2012, minutes before the 2011–2012 budget expired.

The Republican legislators who drew up that amendment claim the amendment was needed to better study the effects of fracking. "We basically said we didn't know [the South Newark Basin] was there before when we did Act 13," said State Sen. Charles T. McIlhinney Jr. (R-Doylestown),[230] who had enthusiastically backed Act 13 but now desperately sponsored the budget amendment when he realized that his comments several months earlier that Act 13 not only protected the environment and local zoning control but would protect his own district from drilling were inaccurate, and that drilling could occur in the affluent Philadelphia suburbs. McIlhinney now said, "We need to slow this down until we can do a study on it—see what's there, see where it is, see how deep it is, study the impact, get the local supervisor's [sic] thoughts on it."[231]

"Where was *our* study?" demanded State Rep. Jesse White (D-Washington County),[232] who actively opposed Act 13 and has been trying to get responsibility on the part of the Industry and the state Legislature regarding the Marcellus Shale drilling. "We were here four months ago [when Act 13 was passed] under the guise of, we had to have uniformity, we had to have consistency, we needed to be fair," said Rep. White, "and now, four months later, we're saying, 'Maybe, for whatever reason, we're going to give a few people a pass.'"[233]

Karen Feridun pointed out, "Studies are not being conducted before drilling begins anywhere else in the state . . . nor are studies being conducted on the potential impacts of the pipeline operations already coming here."[234]

David Meiser, chair of the Bucks County Sierra Club, said the Pennsylvania Legislature "should either exempt all counties from Act 13 and not just try to get special treatment from Sen. McIlhinney's core area, or repeal the law entirely."[235]

Tracy Carluccio, deputy director of the Delaware Riverkeeper Network, called the amendment "outrageous," and accurately noted:

> "This really smacks of cronyism and self-preservation in order to protect a few elite areas in this state that have more power than others . . .
>
> "This is absolutely wrong. This kind of favoritism where a legislator thinks he can protect himself and his turf at the expense of everybody else is what people cannot tolerate and

what they complain about in politics today. We should be making decisions about gas drilling based on sound public policy—not based on what some legislator thinks he can slip in in order to save his own neck."[236]

Sen. McIlhinney proudly claimed the last-minute legislation "makes good on my promise that Act 13 was not intended to apply to Bucks County."[237] And, once again, McIlhinney was wrong. While the toothless moratorium does protect the southeastern counties from drilling, it doesn't protect those counties from drilling-related activities, including construction of pipelines and compressor stations. Significant questions need to be raised why a state law discriminated against the rural counties of the Marcellus Shale while protecting the health and welfare of the more affluent suburban counties that are home for many of the state's most powerful and wealthiest constituents, including the head of the DEP, the lieutenant governor, and the executive director of the Pennsylvania State Association of Township Supervisors.

FOLLOWING THE MONEY INTO NEW YORK

Although New York has a moratorium on fracking until health issues can be fully explored, the natural gas industry and its allies have been seeding the state with political contributions. If Gov. Andrew Cuomo were to allow fracking, he would probably restrict it to the southern tier counties.

During the 2012 elections, the oil/gas industry contributed more than $400,000 to candidates in the 10 southern tier counties, eight of which border Pennsylvania. Most of the contributions were to Republican incumbents from Broome County, likely to be the first county fracked if the state lifts the moratorium.

The industry and its allies contributed $82,428 to Debra A. Preston, the Republican incumbent executive of Broome County, according to data compiled by Common Cause of New York. Oil/gas donations were 22 percent of the $373,858.12 total she had raised.[238] Preston's opponent, Tarik Abdelazim, ran a strong anti-fracking campaign.

In the race in Senate District 52 (Broome, Chenango, and Tioga counties), the industry gave $190,700 to State Sen. Thomas

Libous, an incumbent Republican. With campaign contributions of about $1.3 million, Libous easily outspent and defeated John Orzel, who questioned fracking, and raised only $12,500, and took 33 percent of the vote.[239] Why the industry gave so much to Libous in a race his opponent had almost no hope of winning could be because in addition to being a resident of Binghamton, the county seat of Broome County, Libous is the Senate's deputy majority leader, and someone likely to have a major influence on whether the state goes ahead with plans to frack the southern tier. The senator's official website says one of his top priorities is "fighting for stronger ethical standards in state government."[240]

The oil and gas industry also contributed $47,970 to State Sen. Catharine Young and $28,525 to State Sen. Thomas F. O'Mara.[241] Both Republican incumbents had no opposition, and had been leaders while members of the state Assembly.

In races for Assembly seats in the souther tier counties, the oil/gas industry contributed $45,217 in six races, including three races where the incumbents were unopposed. With one exception, the donations to the winning candidates, five Republicans and a Democrat, were 9.4 to 14.7 percent of their fund-raising totals.[242]

For Congress, Rep. Tom Reed (R-Corning) defeated Nate Shinagawa, who was endorsed by the anti-fracking movement. Reed received $89,521 from individuals and PACs associated with the oil/gas industry.[243]

OBSERVATION

It's possible that significant campaign contributions didn't influence politicians to rush to embrace the natural gas industry and its controversial use of horizontal hydraulic fracking. It's possible these politicians had always believed in fracking, and the natural gas industry was merely contributing to the campaigns of those who believed as they do. However, with the heavy amount of money spent by the natural gas lobby and, apparently, willingly accepted by certain politicians, there is no way to know how they might have voted or how strong their support of fracking would be had there been no lobbying and campaign contributions.

CHAPTER 3
The Pennsylvania Law

Pennsylvania Act 13 of 2012 allows companies to place drilling wells 500 feet from houses, and 1,000 feet from public drinking supplies.[244] Well pads, which are areas where trucks can park and technicians can mix the chemicals for fracking, can be 300 feet from the residential buildings. The law also allows compressor stations to be placed in residential districts 200 feet from a homeowner's property line and 750 feet from houses. Drillers can also place wells 300 feet from streams, creeks, rivers, ponds, and wetlands.

The Pennsylvania law also requires companies to provide fresh water, which can be bottled water, to areas in which they contaminate the water supply, but doesn't require the companies to clean up the pollution or even to track transportation and deposit of contaminated wastewater.

The law doesn't allow for local health and environmental regulation, and forbids municipalities to appeal state decisions about well permits. As a result of numerous concessions, the natural gas industry is given special considerations not given any other business or industry in Pennsylvania.

The law also makes it difficult for anyone to know if there are violations. A CNNMoney investigation, published three months after the law's enactment in February 2012, revealed that the DEP "does not have to notify landowners if a violation is discovered," even if there are significant health, safety, and environmental problems.[245] CNN reporter Erica Fink learned there were 62 safety violations in four years on property owned by four Lycoming County families; there were 26 natural gas wells on their properties. None of the families were notified of the violations. The DEP posts violations online but "the digital records are short on specifics—most importantly whether a violation poses a health risk."[246] When Fink tried to do a hard copy file review, she found additional problems. DEP refused several requests for interviews, said Fink. Additionally, the process to review specific public records "required a visit to the regional DEP office [in Williamsport, Pa.], which had to be scheduled weeks in advance" and the information was "largely in legal and technical language." It was from a meticulous review of paper records that the Fink learned there had been a spill of 294 gallons of frac fluid at one of the wells, but "[t]here was no mention of this spill in DEP's online records, and the paper records did not clearly indicate whether the ground water was tested after the spill."[247]

Numerous sections of Act 13 call for tax waivers or subsidies. Waivers of state sales tax on the purchase of large items or hotel rooms for out-of-state workers were in place before Act 13 was passed. However, the new law also provides subsidies to the natural gas industry, but provides no incentives or tax credits to companies to hire Pennsylvania workers. (The law did not close the "Delaware Loophole," which allows businesses headquartered or incorporated in other states to avoid paying corporate taxes on net income in the state they actually do business in.) The law also allows subsidies for purchase of trucks that weigh at least 14,000 pounds that run on natural gas. The program, which began in December 2012, pays companies $25,000 or 50 percent of the cost, whichever is less, for purchase or to reimburse costs to convert trucks that run on diesel or gasoline. The three year program is expected to cost $20 million.[248]

Gagging the Public and the Health Care Industry

In most states, health care professionals may request specific information, but the company doesn't have to provide that information if it claims it is a trade secret or proprietary information, nor does it have to reveal how the chemicals and gases used in fracking interact with natural compounds. If a company does release information about what is used, health care professionals are bound by a non-disclosure agreement. That agreement not only forbids them from warning the community about water and air pollution that may be caused by fracking, but also forbids them from telling their own patients what the physician believes may have led to their health problems. That section of Pennsylvania's Act 13, as well as similar laws in other states, was drafted by the American Legislative Exchange Council (ALEC),[249] which promotes a right-wing agenda.

In Durango, Colo., Cathy Behr, an emergency room nurse, almost died in August 2008 because of exposure to fracking fluids—and a refusal by a gas driller to release what was in that fluid. According to reporting by Abrahm Lustgarten of *ProPublica*:

> "Behr [was] treating a wildcatter who had been splashed in a fracking fluid spill at a BP natural gas rig. Behr stripped the man and stuffed his clothes into plastic bags while the hospital sounded alarms and locked down the ER. The worker was released. But a few days later Behr lay in critical condition facing multiple organ failure.
>
> "Her doctors searched for details that could save their patient. The substance was a drill stimulation fluid called ZetaFlow, but the only information the rig workers provided was a vague Material Safety Data Sheet, a form required by OSHA. Doctors wanted to know precisely what chemicals make up ZetaFlow and in what concentration. But the MSDS listed that information as proprietary. Behr's doctor learned, weeks later, after Behr had begun to recuperate, what ZetaFlow was made of, but he was sworn to secrecy by the chemical's manufacturer and couldn't even share the information with his patient.
>
> "News of Behr's case spread to New York and Pennsylvania, amplifying the cry for disclosure of drilling fluids. The energy industry braced for a fight.
>
> "'A disclosure to members of the public of detailed information . . . would result in an unconstitutional taking of [Halliburton's] property,' the company told Colorado's Oil and Gas Conservation Commission. . . .
>
> "Then Halliburton fired a major salvo: If lawmakers forced the company to disclose its recipes, the letter stated, it 'will have little choice but to pull its proprietary products out of Colorado.' The company's attorneys warned that if the three big fracking companies left, they would take some $29 billion in future gas-related tax and royalty revenue with them over the next decade.
>
> "In August the industry struck a compromise by agreeing to reveal the chemicals in fracturing fluids to health officials and

regulators—but the agreement applies only to chemicals stored in 50 gallon drums or larger. As a practical matter, drilling workers in Colorado and Wyoming said in interviews that the fluids are often kept in smaller quantities. That means at least some of the ingredients won't be disclosed. . . .

"Asked for comment, Halliburton would only say that its business depended on protecting such information."[250]

The case helped prompt Colorado to rewrite its gas drill regulations to require companies to disclose composition of all fracking fluids. That isn't the case with Pennsylvania.

Drs. Michelle Bamberger and Robert E. Oswald, like most of those in the health and environmental professions, not only call for "full disclosure and testing of air, water, soil, animals, and humans," but also point out that with lax oversight, "the gas drilling boom . . . will remain an uncontrolled health experiment on an enormous scale."[251]

In Pennsylvania, a strict interpretation of Act 13 forbids general practitioners and family practice physicians who sign the non-disclosure agreement and learn the contents of the "trade secrets" from notifying a specialist about the chemicals or compounds, thus delaying medical treatment. The clauses that established that restriction were buried on pages 98 and 99 of the bill.

"I have never seen anything like this in my 37 years of practice," says Dr. Helen Podgainy Bitaxis. She says it's common for physicians, epidemiologists, and others in the health care field to discuss and consult with each other about the possible problems that can affect various populations. Her first priority, she says, "is to diagnose and treat, and to be proactive in preventing harm to others." The new law, she says, not only "hinders preventative measures for our patients, it slows the treatment process by gagging free discussion."

Psychologists are also concerned about the fracking's effects. "We won't know the extent of patients becoming anxious or depressed because of a lack of information about the fracking process and the chemicals used," says Kathryn Vennie of Hawley, Pa., a clinical psychologist for 30 years. She says she is seeing patients "who are seeking support because of the disruption to their environment." Anxiety in the absence of information can produce mental and physical problems, she says.

The law is not only unprecedented, but will "complicate the ability of health departments to collect information that would reveal trends that could help us to protect the public health," says Dr. Jerome Paulson, director of the Mid-Atlantic Center for Children's Health and the Environment at the Children's National Medical Center in Washington, D.C. Dr. Paulson, also professor of pediatrics at George Washington University, calls the law "detrimental to the delivery of personal health care and contradictory to the ethical principles of medicine and public health." Physicians, he says, "have a moral and ethical responsibility to protect the health of the public, and this law precludes us from doing all we can to protect the public." He has called for a moratorium on all drilling until the health effects can be analyzed.

Pennsylvania requires physicians to report to the state instances of 73 specific diseases, most of which are infectious diseases. However, the list also includes cancer, which may have origins not only from chemicals used to create the fissures that yield natural gas, but also in the blowback of elements, including arsenic, present within the fissures. Thus, physicians are faced by conflicting legal and professional considerations.

"The confidentiality agreements are worrisome," says Peter Scheer, a journalist/lawyer who is executive director of the First Amendment Coalition. Physicians who sign the non-disclosure agreements and then disclose the possible risks to protect the community can be sued for breach of contract, and the companies can seek both injunctions and damages, says Scheer.

In pre-trial discovery motions, a company might be required to reveal to the court what it claims are trade secrets and proprietary information, with the court determining if the chemical and gas combinations really are trade secrets or not. The court could also rule that the contract is unenforceable because it is contrary to public policy, which places the health of the public over the rights of an individual company to protect its trade secrets, says Scheer. However, the legal and financial resources of the natural gas corporations are far greater than those of individuals, and they can stall and outspend most legal challenges.

Leaders of the five largest clinical medical societies, in a joint statement published in the *New England Journal of Medicine*, attacked politicians for encroaching upon physician-patient rights and confidentiality in four broad areas, one of which is encompassed in the gag orders related to fracking operations in Colorado, Ohio, Pennsylvania, and Texas. The executives of the the American Academy of Family Physicians, the American Academy of Pediatrics, the American College of Obstetricians and Gynecologists, the American College of Physicians, and the American College of Surgeons, declared:

> "[L]egislators in the United States have been overstepping the proper limits of their role in the health care of Americans to dictate the nature and content of patients' interactions with their physicians. Some recent laws and proposed legislation inappropriately infringe on clinical practice and patient–physician relationships, crossing traditional boundaries and intruding into the realm of medical professionalism. . . .
>
> "By reducing health care decisions to a series of mandates, lawmakers devalue the patient–physician relationship. Legislators, regrettably, often propose new laws or regulations for political or other reasons unrelated to the scientific evidence and counter to the health care needs of patients."[252]

Dr. Eli Avila, Pennsylvania's health secretary, dismissed arguments that the new law restricts physicians. Dr. Avila, who resigned in September 2012, had claimed that physicians could share information with patients and other medical providers, as well as to report health concerns to the state.[253] However, no matter what Dr. Avila claimed, he was wrong; Act 13, in language likely to be so interpreted within the courts, specifically restricts physicians from disclosing corporate "trade secrets" to patients and other medical providers.

House Speaker Sam Smith (R-Punxsutawney), responding to criticism by physicians, health professionals, and others about the impact of Act 13, called their complaints "outrageous" and "irresponsible."[254] He also claimed, "Pennsylvania has the most progressive hydraulic fracturing disclosure law in the nation. It is designed for transparency and access, and it provides unfettered access to physicians or other medical professionals who need information to treat their patients."

Many physicians disagree. Dr. Alfonso Rodriguez, a nephrologist from Dallas, Pa., sued the Pennsylvania DEP secretary, attorney general, and the chair of the Public Utilities Commission. He filed the suit after treating a trucker who was splashed by fracking fluid. That trucker now suffers from kidney failure and is on dialysis. Dr. Rodriguez says he had tried to find out the chemical ingredients of that fluid:

> "That's when we ran up against the firewall. They keep telling me the stuff's benign, just soap and emulsifiers. And I say, 'Just send me the sheet.' They say, 'We can't do that. It's not allowed.'"[255]

The complaint charges: "the Medical Gag Rule is an unconstitutionally overbroad content-based regulation of speech [that requires] a confidentiality agreement as a condition precedent to receive information needed for the ethical and competent treatment of a patient in an emergency situation to prohibit communications which are not narrowly tailored to advance a compelling governmental interest."[256] He says the gag rule, which he believes violates his 1st and 14th Amendment rights, requires him to either violate a non-disclosure agreement or his oath as a physician. As a physician, says Dr. Rodriguez, he is "ethically required to secure any information necessary to provide competent treatment to his patients. In order to secure needed information from gas drillers and/or their agents and/or vendors in emergency situations, plaintiff is required by the Medical Gag Rule, upon request, to waive rights secured to him."[257]

Although Pennsylvania is determined to protect the natural gas industry, not everyone in the industry agrees with the need for secrecy. Dave McCurdy, president of the American Gas Association, says he supports disclosing the contents included in fracturing fluids. In an opinion column published in the *Denver Post*, McCurdy further argued, "We need to do more as an industry to engage in a transparent and fact-based public dialogue on shale gas development."[258]

The Natural Gas committee of the U.S. Department of Energy agrees. "Our most important recommendations were for more transparency and dissemination of information about shale gas operations, including full disclosure of chemicals and additives that are being used," said Dr. Mark Zoback, professor of geophysics at Stanford University and a Board member.[259] Both McCurdy's statement and the Department of Energy's strong recommendation about full disclosure were known to the Pennsylvania General Assembly when it created the law that restricted health care professionals from disseminating certain information that could help reduce significant health and environmental problems from fracking operations.

Constitutional Issues

Two separate New York courts have upheld the rights of local governments to create and enforce zoning ordinances [*Cooperstown Holstein Corporation v. Town of Middlefield*[260] and *Anschutz Exploration Corp v. Town of Dryden*[261]]. Both rulings came in the same week in February 2012. But several states have taken preemptory rights over local ordinances.

Following a presentation by NOAA chemist Dr. Steven Brown, Erie, Colo.—primary site of a NOAA research study that found high atmospheric levels of toxins and carcinogens—put a moratorium on fracking within the town limits.[262]

Longmont, Colo., an 86,000 population city 10 miles north of Erie, also put a moratorium on oil and gas permits within residential areas of the city.[263] Then, in July 2012, the Longmont city council passed an ordinance that continuing the moratorium was necessary "for the immediate preservation of the public peace, health, or safety." The ordinance was overturned by Gov. John Hickenlooper who declared that allowing the ordinance to stand would "stir up a hornet's nest" to encourage other towns to also take local control of their zoning.[264] Later that month, the state's attorney general sued Longmont on charges it usurped the power of the state. The city's residents were undeterred. Led by Michael Bellmont, a musician and insurance agent who said he drew inspiration from the ban on fracking passed in 2010 by the Pittsburgh, Pa., city council,[265] the citizens easily got enough signatures to force a ballot initiative onto the November ballot to amend the city charter to ban all fracking and waste storage. Against a $500,000 advertising onslaught by the natural gas industry,[266] and opposition from local and regional newspapers, the citizens approved that initiative.[267] A month later, Gov. Hickenlooper announced the state would not pursue legal action against Longmont because "there is uncertainty on whether [Colorado] has legal standing to sue, because nothing has been taken from the state."[268] However, he also said that the citizen action is "a clear taking from oil and gas mineral rights owners," and that the state would "help and support" any suit from citizens or the gas industry against the city.[269]

In Pennsylvania, Act 13 guts local governments' rights of zoning and long-term planning, doesn't allow local governments to impose restrictions on gas well drilling stricter than the state law, and doesn't allow municipalities to deny any gas company the state-mandated right to drill. The state legislature specifically inserted those clauses into Act 13 to negate a Pennsylvania Supreme Court decision in 2009 that affirmed the right of municipalities to determine zoning and planning regulations. [*Huntley Huntley v. Oakmont* 600 Pa. 207, 964 A.2d 855[270]].

The law also gives the Public Utilities Commission (PUC) the right to overturn local zoning, and to act as a *de facto* court of appeals for individuals and the gas companies.[271] The PUC is composed of appointed officials and, thus, is likely to have a predetermined political bias.

In April 2012, seven municipalities (Cecil, Mount Pleasant, Nockamixon, Peters, Robinson, and South Fayette townships, the borough of Yardley), the Delaware Riverkeeper Network, and physician Mehernosh Khan filed suit against the Commonwealth to block enforcement of parts of the law that revoked local government authority to create and enforce zoning ordinances, and charged that Act 13 was unconstitutional under several sections of the state's constitution and the 14th Amendment of the U.S. Constitution. The suit charged that Act 13 was created "to elevate the interests of out-of-state oil and gas companies," and that by overriding local zoning authority, "municipalities can expect hundreds of wells, numerous impoundments, miles of pipelines, several compressor and processing plants, all within [their] borders, they will be left to plan around rather than for orderly growth."

According to the 117-page suit:

Through Act 13, the General Assembly has mandated that Municipal Petitioners must:
  a. modify their zoning laws in a manner that fail to give

61

consideration to the character of the municipality, the needs of its citizens and the suitabilities and special nature of particular parts of the municipality; 53 P.S.§ 10603(a).

b. modify their zoning laws in a manner that would violate and contradict the goals and objectives of Petitioners' comprehensive plans; 53 P.S. §10605.

c. modify zoning laws and create zoning districts that violate Petitioners' constitutional duties to only enact zoning ordinances that protect the health, safety, morals and welfare of the community; See, 53 P.S. §10604.

d. conduct Public Hearings to gather citizen comments regarding authorized oil and gas development in residential and commercial districts as a permitted use by right even though such comments and evidence cannot be considered by Petitioners who, by state law, must approve the state's zoning scheme regardless of the findings of the elected officials in violation of 53 P.S. §10908.

e. conduct Public Hearings negating citizens' due process rights to meaningful participation in proceedings involving the adoption of a zoning ordinance; Messina v. East Penn Twp., 995 A.2d 517 (Pa. Ct. 2010).

f. pass zoning laws without affording its citizens due process that will result in the zoning laws being void ad initio; Luke v. Cafaldi, 932 A.2d 45 (Pa. 2007).

g. allow heavy industrial uses in all zoning districts, including residential areas, near homes, schools, churches and nursing homes in violation of 53 P.S. §10605.

h. must enact zoning laws that do not allow for the orderly development of their respective communities; and, See, 53 P.S. §10605.

i. adopt zoning laws that are an improper use of the sovereign's police powers in violation of the U.S. Constitution and Pennsylvania Constitution.

In the Petition for Review, Petitioners assert that:

a. Act 13 violates Article I, Section 1 of the Constitution and Section 1 of the 14th Amendment to the United States Constitution as Act 13's zoning scheme is an improper exercise of the police power that is not designed to protect the health, safety, morals and public welfare of the citizens of Pennsylvania.

b. Act 13 violates Article I, Section 1 of the Constitution because it allows for incompatible uses in like zoning districts in derogation of municipalities' comprehensive zoning plans and therefore constitutes an unconstitutional use of zoning districts.

c. Act 13 violates Article I, Section 1 of the Constitution as Act 13's allowance of oil and gas development activities as a permitted use by right in every zoning district renders it impossible for municipalities to create new or to follow existing comprehensive plans, zoning ordinances or zoning districts that protect the health, safety, morals and welfare of citizens and to provide for orderly development of the community in violation of the MPC resulting in an improper use of its police power;

d. Act 13 violates Article Section 32 of the Constitution because Act 13 is a "special law" that treats local governments differently and was enacted for the sole and unique benefit of the oil and gas industry;

e. Act 13 is an unconstitutional taking for a private purpose and an improper exercise of the eminent domain power in violation of Article I Sections 1 and 10 of the Constitution.

f. Act 13 violates Article I, Section 27 of the Constitution by denying municipalities the ability to carry out their constitutional obligation to protect public natural resources, removing muncipalities' constitutionally mandated duty to protect the environment;

g. Act 13 violates the doctrine of Separation of Powers because, through its provision that allows for advisory opinions, Act 13 permits an Executive agency, the Public Utility Commission, to play an integral role in the exclusively Legislative function of drafting legislation;

h. Act 13 violates the doctrine of Separation of Powers because it entrusts an Executive agency, the Public Utility Commission with the power to render opinions regarding the constitutionality of Legislative enactments, infringing on a judicial function. See, *Commonwealth v. Allshouse*, 33 A.3d 31 (Pa. Super. 2011);

i. Act 13 unconstitutionally delegates power to the Department of Environmental Protection without any definitive standards or authorizing language.

j. Act 13 is unconstitutionally vague because its setback provisions and requirements for municipalities fail to provide the necessary information regarding what actions of a municipality are prohibited.

k. Act 13 is unconstitutionally vague because its timing and permitting requirements for municipalities fail to provide the necessary information regarding what actions of a municipality are prohibited. 8 l.

l. Act 13 is an unconstitutional "special law" in violation of Article Ill, Section 32 of the Constitution which restricts health professionals' ability to disclose critical diagnostic information when dealing solely with information deemed proprietary by the natural gas industry.

m. Act 13's restriction on health professionals' ability to disclose critical diagnostic information is an unconstitutional violation of the single- subject rule enunciated in Article Section 3 of the Constitution.[272]

By mixing residential living with industrial activity, "The state has surrendered over 2,000 municipalities to the industry," said Ben Price of the Community Environmental Legal Defense Fund. He called the law "a complete capitulation of the rights of the people and their right to self-government."[273]

In securing support for Act 13, the Corbett Administration had claimed the support of the Pennsylvania State Association of Township Supervisors (PSATS), an association of 1,455 local municipalities and townships. "PSATS has been cited again and again by the Governor and the Majority leadership to show how their legislation had the 'participation' and 'support' of local elected officials," *Marcellus Monthly* reported.[274] However, it was not the members who supported Act 13 but PSATS' paid executives, a split that become apparent with the subsequent suit.

David Ball, councilman for Peters Twp. in the southwestern part of the state, charged the association's reporting of the suit in its April 2012 newsletter as being "deliberately misleading." In a letter dated May 7, 2012, to David Sanko, PSATS executive director, Ball attacked Sanko's failure to defend the townships:

"Where was PSATS whose job it is to advocate to ensure our protections? The court confirmed with its injunction that the legislation posed real danger to municipalities. By its silence, PSATS exposed municipalities to danger that needed to be enjoined. . . .

"PSATS gives all appearances of having sold out to the gas industry. During legislative debate you failed to accurately inform municipalities of the utter destruction of their zoning capabilities that Act 13 (or then HB 1950) represented and you produced at least two documents that legislators were able to hide behind and claim that 'PSATS endorses this bill'. You denied that you endorsed the bill on several occasions in meetings with concerned local officials but then did nothing to clearly state that publically. You did nothing to solicit input from potentially impacted townships; neither did many of our legislators, something you should have been all over. You were made aware of many problems with the legislation by local officials who took the time to come to meet with you and the executive committee and did nothing to convince legislators of those short comings. When the State government takes away the right of local officials to plan and zone, it takes away their ability to govern. When you do not react to this, you are in direct conflict with [Association bylaws] Section B. [To protect the township against any attempt to abolish it as a governmental unit.] By not responding to the usurping of constitutional rights to the protection of safety, health and welfare and the vesting of those rights and duties in the state government, you violate section C. [To resist any effort at further centralization of governmental powers by depriving the township of any rights, duties or privileges that it now possesses.] Allowing the State to pass a bill like HB 1950 without an all out fight is blatant dereliction of duty with respect to Section F. [To secure from the General Assembly legislation that will enable township government to function more efficiently.] Your commentary in the PSATS Bulletin serves only industry's and the Governor's interests and perspectives. When will PSATS begin to react in a meaningful way to real and vital issues that have the potential to emasculate local government? This is what we pay you for."

In Berks County, several citizens went to municipal meetings to speak against Act 13 and to advise the local governments of the impact the Act had against their communities. However, PSATS "was actively working against us," says Karen Feridun of Berks Gas Truth. She says PSATS "was telling municipalities that it was a waste of their time and that it was a state, not a local, issue."

At the PSATS annual meeting in May 2012, with Gov. Tom Corbett as the featured guest speaker, the membership passed two resolutions:

12-36 RESOLVED, That PSATS oppose any legislation that would remove, reduce, or inhibit local government authority or existing local subdivision, land use, and zoning controls.

12-37 RESOLVED, that PSATS oppose any legislation that would pre-empt the existing authority of townships to regulate land use through zoning and subdivision and land development ordinances, including any amendments to the Coal and Non-Coal Surface Mining and Conservation Act and Oil and Gas Act.

In a subsequent email to Sanko, dated June 7, 2012, David Ball again addressed the issue of what he and others saw as the Association's non-compliance of member wishes:

"Yesterday in Commonwealth Court one of the attorneys for the Respondents [the Corbett Administration] stated directly that 'PSATS supports the zoning provisions of this ACT'.

"I, and the supervisors of a lot of other townships, request that PSATS issue an immediate and unequivocal denial that it supports these provisions and cite the resolutions passed at the convention. . . ."

In response to that email, Sanko claimed that if Chapter 33, the portion addressing zoning, "had been a standalone bill, PSATS would have opposed it. . . . But Act 13 was a comprehensive piece of legislation and the zoning provisions were but one component that affected PSATS' membership." Deliberately obfuscating the effect of Act 13 upon municipalities, Sanko claimed "Act 13, as a whole, was the best product that the General Assembly had put forth to date . . ."

By speaking on behalf of Act 13, while staying silent on certain parts that allowed the state to take jurisdiction over local zoning in order to make gas drilling easier for energy companies, PSATS' professional staff abrogated the rights of municipalities to protect the health, safety, and welfare of residents, as was brought out in the legal challenge to the entire Act. Like David Ball, many of supervisors and council members who oppose Act 13 are conservative Republicans. They don't object to the drilling, as long as public health and safety are protected. But, they do object when the state usurps local authority in order to benefit one industry.

The Commonwealth Court ruled against the state, and upheld the rights of municipal government to establish adequate zoning, specifically agreeing with the petitioners on Counts 1, 2, 3, and 8 of their suit. In his majority opinion issued in July 2012, President Judge Dan Pellegrini, wrote that the zoning provision of Act 13:

> "violates substantive due process because it does not protect the interests of neighboring property owners from harm, alters the character of neighborhoods and makes irrational classifications— irrational because it requires municipalities to allow all zones, drilling operations and impoundments, gas compressor stations, storage and use of explosives in all zoning districts, and applies industrial criteria to restrictions on height of structures, screening and fencing, lighting and noise."[275]

In a separate order, Judge Pellegrini declared the zoning provision of Act 13 "unconstitutional and null and void." As expected, Gov. Tom Corbett appealed the decision to the state Supreme Court. In a public statement, Corbett continued to hammer on the economic benefits of natural gas drilling, while ignoring the health and pollution effects from fracking. In a subsequent ruling, Judge Pellegrini informed the Corbett administration, "Jobs do not justify the violation of the constitution."[276] Corbett also specifically noted full support of the Pennsylvania State Association of Township Supervisors. "It's frustrating," says Dave Ball, "that even after the membership passed the two resolutions that strongly opposed the zoning provisions of Act 13 that executives of PSATS have refused to publically announce those votes, and that the Governor [who knows of those resolutions] continues to claim township support."

The petitioners filed a cross-complaint to the Supreme Court, which would allow that Court not only to uphold the lower court's decision but also to rule on the other eight complaints. The state filed a 45-page brief, arguing that the Commonwealth Court "failed to acknowledge and uphold the supreme authority of the Legislature."[277] It was an unusual position since the Constitution, not what the legislature does, is the supreme authority. To overturn the Commonwealth Court's ruling, the Supreme Court would have to declare its own recent decision in *Huntley, Huntley v. Oakmont* to be null and void, a highly unlikely possibility.

Frustrated by PSATS' failure to defend municipalities while allowing the Corbett administration to continue to claim PSATS support, members had begun to suggest they might not renew their membership. Finally, at the end of August 2012, more than three months after members passed resolutions opposing parts of Act 13, and more than a month after the court determined that part of Act 13 was unconstitutional, David Sanko finally issued a lame statement. "Now that the courts have made their determination, we're standing with that," the PSATS executive director said, and now claimed that Act 13 "was a compromise for everybody [and] there was something ugly in that bill for everyone."[278]

The arrogance of PSATS executives, who now reluctantly supported the will of the membership, was evident in its impudent claim, and abrupt dismissal of requests for information about the resolutions and detailed requests for the organization's budget, much of it from public tax funds, and communications between it and the governor's office that might suggest collusion. PSATS claimed it was a private organization, and not bound by public records disclosures. In September 2012, the state's Office of Open Records ruled in *Brasch v. PSATS [Docket 2012-1184]* that PSATS "has not met its burden of proof to withhold responsive records," and that the organization, created by statute, "exists solely as an extension of, and to serve, township governments in the Commonwealth that are otherwise subject to the RTKL [right-to-know law]."

The state's PUC also supporting the Corbett administration decided that the ruling of the Commonwealth Court that vacated parts of Act 13 apparently didn't apply to it. The PUC withheld payments of about $1 million from drilling fees to Cecil, Mt. Pleasant, Robinson, and South Fayette townships, which were part of the suit against the Corbett administration and the PUC. State Rep. Jesse White called the decision to withhold funds "political extortion."[279] The Commonwealth Court agreed. In October 2012, Senior Judge Keith Quigley issued a "cease and desist" order to the PUC, ruling that the agency did not have authority to determine local zoning ordinances or to withhold drilling impact fees.[280]

CHAPTER 4
The Economics of Fracking

It takes about $5 million to bring a well into production, says Patrick Creighton of the industry-sponsored Marcellus Shale Coalition.[281] In 2011, the natural gas industry in Pennsylvania produced about one trillion cubic feet of natural gas, and had revenues of about $3.5 billion for its wells in the Marcellus Shale.[282]

However, economist Deborah Rogers says "only 20 percent of wells drilled will actually make money," and notes:

> "Eighty percent can easily be uneconomic. That is a whole lot of land used up in a search for 20 percent of the wells that will make money. Eighty-five percent of wells are abandoned in the first five years. And seven years is the average life of a well, rather than the 30 promised by industry."[283]

That projection is considerably less than the 100 years that President Obama had said could be the life of the natural gas industry in the shales that could serve as the "transition" fuel to wind, solar, and water energy. Nevertheless, under the Obama administration, the U.S. has increased its development of alternative green energy, but has run into significant resistance from drillers and distributors of the fossil fuel.

The Lure of Jobs

The natural gas industry claims that drilling in the Marcellus Shale will bring several hundred thousand jobs in a depressed economy, and will help the U.S. reduce its oil dependence on foreign nations. The U.S. Chamber of Commerce and the Pennsylvania Chamber of Commerce, noted for supporting business and conservative candidates for political office, have strongly promoted fracking to extract natural gas.[284] What they aren't admitting is that most of the jobs are temporary, the work conditions mentally and physically challenging, and employers are not giving their employees health benefits. In July 2012, the Chambers announced a multi-million dollar "advocacy and education campaign" to lobby Pennsylvania, West Virginia, Ohio, and other states with large shale deposits about what they see as the benefits of fracking and natural gas drilling.[285] The U.S. Chamber, which spends more in lobbying expenses than any company or organization, spent about $901.2 million between 1998 and 2012,[286] with $95.7 million of it spent in 2012.[287]

President Barack Obama specifically noted that the development of natural gas as an energy source to replace fossil fuels could generate 600,000 jobs, a claim that had previously been advanced by the natural gas industry. "[T]he industry is working hard to move forward responsibly and will continue to create good paying, family sustaining jobs that will make our economy more competitive," said Gene Barr, president of the Pennsylvania Chamber.[288] However, the Chamber and the Pennsylvania politicians who pushed fracking as a way to increase jobs for Pennsylvanians don't acknowledge that even a most cursory look at the pick-up trucks and private vehicles behind the gates of any well pad often reveals more Texas and Oklahoma license plates than those of Pennsylvania.

In addition, thousands of vendors and subcontractors to the natural gas industry provide myriad products, including signs, telecommunication devices, guard shacks, hard hats, parts for trucks, and compressors.

The Marcellus Shale Coalition claimed 88,000 new jobs were created in Pennsylvania in 2010 because of Marcellus Shale drilling. [289] In an OpEd column for the *Philadelphia Inquirer* in September 2012, Kathryn Z. Klaber, Marcellus Shale Coalition president, argued, "Marcellus development is creating tens of thousands of jobs at a time when they're most needed," and claimed that according to "state data, nearly 239,000 jobs across the commonwealth are tied to our industry."[290]

However, the claims of significant job creation are inflated. "Those who lobby [for fracking] tend to overstate the benefits and understate the costs," said Pennsylvania Rep. Greg Vitali (D-DelawareCounty).[291] Research studies by economists Deborah Rogers, Dr. Jannette M. Barth, and others agree with Vitali and debunk the idea of significant job creation.

Rogers used data provided by the Department of Labor's Bureau of Labor Statistics to discredit the industry's inflated jobs creation estimates. The Perryman Group, commissioned by the U.S. Chamber of Commerce, claimed there were 111,131 jobs created by drillers in northern Texas during 2008.[292] However, BLS statistics reveal there were only 166,500 jobs created in the entire United States in the production sector that year. Chesapeake Energy claimed 53,200 jobs for the Fort Worth/Arlington area in 2010;[293] Rogers questioned that claim since there were only 93,800 jobs created in the production sector for the entire country in 2010.

"The grandiose job creation claimed by the industry is not at all consistent with data from unbiased, publicly available sources," said Dr. Barth in testimony before the New York State Committee on Environmental Conservation in October 2011.[294]

However, Dr. Barth testified, "Publicly available Pennsylvania data available at that time clearly showed that total job creation in the entire state was only 65,600. And half of these jobs were in 'education and health' and in 'leisure and hospitality.' The Keystone Research Center reported that job creation was only 5,669 net jobs between the fourth quarter of 2007 and the fourth quarter of 2010.[295] "The gas industry," said Dr. Barth, "has an incentive to mislead the public in order to gain public support for gas drilling."[296]

No matter how few or how many jobs are created, drilling for natural gas in the Marcellus Shale is not because energy companies are good citizens who want to improve the local economy and hire a robust work force. The companies have every intention of showing a profit while giving their investors a good return on investment. However, that investment is tainted by how the energy exploration companies received billions of dollars in loans from Wall Street financial institutions. The *New York Times* in October 2012 reported:

> "After the [mortgage] financial crisis, the natural gas rush was one of the few major profit centers for Wall Street deal makers, who found willing takers among energy companies and foreign financial investors.
> "Big companies like Chesapeake and lesser-known outfits

like Quicksilver Resources and Exco Resources were able to supercharge their growth with the global financing, transforming the face of energy in this country. In all, the top 50 oil and gas companies raised and spent an annual average of $126 billion over the last six years on drilling, land acquisiion and other capital costs within the United States, double their capital spending as of 2005, according to an analysis by Ernst & Young."[297]

The Lure of Immediate Gratification

Because the oil and gas culture are a part of the lifestyle of the people of Pennsylvania, Texas, Oklahoma, and other states, "people have fully embraced fracking [to gain immediate financial benefit] without knowing its impact," says Elisabeth Radow, an attorney specializing in real estate law, financing, construction, and environmental conservation

When the natural gas industry first came into the Marcellus Shale to lease land, the sales people held out the lure of a lump sum payment plus a royalty on gas extracted. For families faced by the recession, leasing mineral rights for unused land or converting portions of non-producing farm land to the corporations might have been a way to get out of debt and even have a decent bank account to prepare for future emergencies. Some owners with dozens of acres to lease became rich; most didn't.

During the first years of exploration in the Marcellus Shale, energy companies paid as little as $5–$25 an acre for a five year lease, plus a 12.5 percent royalty. By 2009 bonuses increased to $1,000–$7,000 and royalties sometimes exceeded the legal minimum of 12.5 percent as landowners became more knowledgeable, says Dave Messersmith of the Penn State Extension office. Landowners in the Marcellus Shale occasionally negotiated additional benefits, including clean-up costs and a certain amount of free natural gas for home heating use. However, there are several provisions that do not benefit landowners. Energy companies can also tie up land indefinitely, even with non-producing wells, if there is no expiration date; contracts can specify the royalties are only paid for working wells on a property. As landowners became more financially savvy, they demanded specific expiration dates. The rush to grab as much land as possible, and to show activity, was noticed as early as May 2010 when Marcus C. Rowland, Chesapeake's chief financial officer, acknowledged, "at least half and probably two-thirds or three quarters of our gas drilling is what I would call involuntary."[298]

By 2012, the energy companies "took all the land they needed" in Pennsylvania, says Messersmith, "so now they're buying just smaller patches."

In Pennsylvania, most mineral rights are owned by few people. Research by Dr. Timothy W. Kelsey, Dr. Alex Metcalf, and Rodrigo Salcedo for the Center for Economic & Community Development, reveals:

> "[O]wnership of the land in the Pennsylvania counties with the most Marcellus drilling activity is concentrated in a relatively small share of residents, and in owners from outside the county. The majority of residents of these counties together own little of the total land area, and so have relatively little 'voice' in the critical leasing decisions which affect whether and how Marcellus shale drilling will occur in their county. Half of the resident landowners in the counties together only control 1.1 percent of the land area, and renters have no 'voice' at all. Rather it is the top 10 percent of resident landowners, plus outside landowners (both public and private), who are able to make the major leasing decisions that affect the rest of the community. . . .
>
> "[A] majority of lease and royalty income from Marcellus shale development will go to a relatively small share of the resident population in these counties, with much of the remainder going to others outside the counties. A little less than half (48.9 percent) of the lease and royalty dollars in these counties will go to the top ten percent of local landowners, while 39.8 percent will go to the public sector or nonresident landowners. The remaining 11.3 percent of lease and royalty income will be divided between the bottom 90 percent of local landowners."[299]

In addition to the lack of voice and income, there are additional realities not addressed by proponents of fracking. In what the industry derisively calls the "Sacrifice Zone," many homeowners can't get second mortgages, home equity loans, or insurance, and their houses are being devalued.

"Even before the drilling commences, many upstate New York homeowners with gas leases cannot obtain mortgages," says Elisabeth Radow. Among major lending institutions that have refused to issue mortgages are Bank of America and Wells Fargo, says Radow.

The North Carolina State Employees Credit Union issued a statement in April 2012 that further clarified the reality about purchasing houses from owners who have already leased mineral rights or from future owners who have secured both surface and mineral rights but plan to lease rights, hoping the royalties would offset the mortgage payments:

> "The standard Fannie Mae/Freddie Mac Deed of Trust docu-
> ment recorded for most real estate liens prohibits the home-
> owner from selling or transferring any part of the property
> during the term of the loan without obtaining prior written
> approval from an official of the financial institution holding the
> mortgage. This includes the oil, gas and minerals found on the
> property. Any property financed with a State Employees'
> Credit Union mortgage falls under the aforementioned restrict-
> tion. Approval of exceptions from State Employees' Credit Union
> would not be granted due to heightened risk concerns associ-
> ated with extraction of these natural resources, including
> hydraulic fracturing technology (otherwise known as fracking
> or horizontal drilling)."[300]

Homeowners are also finding it difficult to get insurance. Nationwide Insurance no longer insures homes and property in the heavily industrialized areas of natural gas drilling. A company news release in July 2012, clearly states:

> "Fracking-related losses have never been a covered loss
> under personal or commercial lines policies. Nationwide's
> personal and commercial lines insurance policies were not
> designed to provide coverage for any fracking-related risks . . .
> "From an underwriting standpoint, we do not have a comfort
> level with the unique risks associated with the fracking process
> to provide coverage at a reasonable price."[301]

What they're saying "is they won't cancel your insurance but if there's drilling on your property, they won't renew the insurance next year," says Radow. Other insurance companies are likely to not insure or renew policies of homeowners who allow fracking on their property.

Under federal law, persons living in a flood plain are required to have flood insurance.[302] If an insured house is flooded and uninhabitable, the Federal Emergency Management Agency (FEMA) could buy that house at fair market value and then forbid any construction on the site. However, FEMA is rejecting buy-outs for homeowners who have signed leases to natural gas companies. A review of public documents by Sharon Kelly of DeSmogblog.com revealed "at least 18 homeowners in Pennsylvania have been denied access to the Hazard Mitigation program because of oil and gas leases or pipeline rights-of-ways on their properties."[303] Part of the reason is that by signing away mineral rights, "The landowner and the oil company now co-own the rights to use the surface of the property and the oil company owns all of the oil and gas under the property," FEMA attorney Michael C. Hill wrote to the Pennsylvania Emergency Management Agency in June 2011.[304] "Thus the landowner alone cannot transfer all of the surface rights nor the oil and gas rights," he pointed out. If FEMA does not buy the affected property, the consequences can be significant, says Kelly:

> "If their land is not acquired, the affected landowners may be caught in a bind—unable to sell their land, they may face sharply higher federal flood insurance rates, which are hiked after repeated flooding, partly as an incentive for homeowners to participate in the HGMP [Hazard Mitigation Grant Program]. If FEMA moves to acquire the land, the government may be legally required to conduct a review under the National Environmental Policy Act, some legal experts say, but this process is famously slow and can stretch on for decades.[305]

Jennifer Canfield, a former real estate agent in Wayne County, Pa., has been unable to sell her own house on six acres near leased property. "Nobody looks at it," she says, "because people don't want to be anywhere near the drilling. If the economy turned around, we still wouldn't get these people back."[306]

"In general, homes that are in close proximity of fracking may have a decreased home value of about 20 percent as a result of construction, noise, traffic, airborne dust, and road damage," according to the 360 Mortgage Group in Austin, Texas.[307] In many cases, the value drops by as much as half; in some cases, it's nearly impossible to sell houses near natural gas wells.

"The most important thing to many people is their homes," says Radow, who explains that owning a home "is an emotional safe place." But when contaminated water makes it unsafe, "and fracking makes it impossible for you to sell your home or that you may only get a fraction of its value, people become upset," Radow says.

Richard Plunz, director of the Urban Design Lab at Columbia University, told *Columbia Magazine*:

> "If there's a lease on your land, your property is devalued. People didn't understand that initially. They were told by the gas-company landsmen, 'You're going to make a fortune, and you won't even see a well.' But even without a well, nobody is going to buy property that has a lease. The value of the neighbors' property probably decreases, too. No one wants to buy a house in an industrializing landscape.
>
> "The long-term economic prospects for these towns are diminished. The land will be undesirable, scarred with roads and well pads and possibly contaminated. The owners will have collected their proceeds from the production as long as possible, but when the profits end they can simply walk away. With that, the town's tax revenue fades."[308]

Seizing Private Property

EMINENT DOMAIN

Pennsylvania's Act 13, written by conservatives who traditionally object to "Big Government" and say they are the defenders of individual property rights, allows any corporation involved in the natural gas industry to "appropriate an interest in real property [for] injection, storage and removal" of natural gas. Thus, homeowners and developers who don't have mineral rights to their property may find themselves forced to surrender land at sale prices that may not be as high as market value to natural gas companies that wish to run pipes both beneath and on top of the land.

Under the Constitution, the government may seize private property if that property must be taken for public use and the owner is given fair compensation. Although the exercise of eminent domain to seize land for the public good is commonly believed to be restricted to the government, federal law[309] permits natural gas companies to use it.

Sandra McDaniel, of Clearville, Pa., says she was forced to lease five of her 154 acres to Spectra Energy Corp., which planned to build a drilling pad. "The government took it away, and they have destroyed it," she said.[310]

In Tyrone Twp., Mich., Debora Hense returned from work in August 2012 to find that Enbridge workers had created a 200 yard path on her property and destroyed 80 trees in order to run a pipeline. Because of an easement created in 1968 next to Hense's property, Joe Martucci of Enbridge Energy Partners said his company had a legal right to "to use property adjacent to the pipeline." Martucci says his company offered Hense $40,000 prior to tearing up her land, but she refused. Hense says she had a legal document to prevent Enbridge from destroying her property; Enbridge says it had permission from the Michigan Public Service Commission.[311]

Three recent Texas cases underscore the power governments have allowed energy companies to develop. Chuck Paul, who lost about 30 of his 64 acre farm because of required easements by the natural gas industry, told the *Fort Worth Weekly*:

> "The gas companies pay a one-time fee for your land, but you lose the right to utilize it as anything more than grassland forever. . . . You can never build on those easements. They took my retirement away by eminent domain.
>
> "They just screwed me. And they might want to come back and take more. And there's nothing you can do to stop them. If they'd have paid me fair market value, I would have sold it to them. But with the right of eminent domain they have, well, they can do anything they want."[312]

Julia Trigg Crawford of Direct, Texas, lost a 50-foot strip of pasture to TransCanada, which is building the controversial Keystone XL Pipeline. Crawford had rejected TransCanada's offer of $10,395 for the easement, leading to the seizure and subsequent case decided in the company's favor by the Lamar County Court-at-Law in August 2012.[313]

In Arlington, Texas, Ranjana Bhandari and her husband, Kaushik De, refused to grant Chesapeake Energy the right to take gas beneath their home, although Chesapeake promised several thousand dollars in payments. "We decided not to sign because we didn't think it was safe, but the Railroad Commission doesn't seem to care about whose property is taken," Bhandari said.[314] The Railroad Commission regulates oil and gas in Texas. Its response was that not only did it consider fracking to be safe, but that "We are charged by the legislature to make sure hydrocarbons don't stay underground and go to waste." Chesapeake seized the mineral rights and will capture natural gas beneath the family's homes; under Texas law, the nation's second largest energy company has no obligation to pay anything to the landowners. Between January 2005 and October 2012, the Railroad Commission approved all but five of Chesapeake's 1,628 requests to seize mineral rights, according to the Reuters investigation.

Texas law allows a common carrier to seize land with minimal justification. The Texas Supreme Court, in *Texas Rice Land Partners and Mike Latta v. Denbury Green Pipeline–Texas,*[315] had previously ruled, "Even when the Legislature grants certain private entities 'the right and power of eminent domain,' the overarching constitutional rule controls: no taking of property for private use." In that same opinion, the Court also ruled, "A private enterprise cannot acquire unchallenged-able condemnation power . . . merely by checking boxes on a one-page form and self-declaring its common-carrier status." However, Texas has no public agency to set standards for seizing property by eminent domain; this lack of standards and an agency to enforce them allows the practice that took land from Chuck Paul, Julia Trigg Crawford, Ranjana Bhandari, Kaushik De, and thousands of others to continue.

FORCED POOLING

Related to Eminent Domain is Forced Pooling, which requires landowners to surrender mineral rights to private energy companies. Thirty-nine states allow forced pooling.

The concept of forced pooling, dating to the 1930s, allows energy companies to ask the state to allow it to "pool" mineral rights on large areas (often at least 640 acres, depending upon the state), and only when a majority of landowners in the projected parcel (usually 65–90 percent) have already signed over their mineral rights. Forced pooling, essentially, is a threat not to owners of larger properties who may object and not sign leases, but to the owners of smaller properties, sometimes as little as a quarter acre, who are in the larger parcel.

Under all state laws, energy companies must notify landowners in the prospective parcel; the landowners have a right to protest at a public hearing. If the state approves, then all landowners holding mineral rights are forced to sell or lease their rights to the energy company.

Forced pooling proponents say that it is an improvement upon the "Rule of Capture" that allows energy companies with wells on adjacent properties to "capture" the oil or gas beneath the property that is not owned or leased by the company. Forced pooling, say the energy companies, allows them to be more efficient in natural gas extraction by preventing a patchwork of land that is leased and unleased, reduce the number of wells, and benefit all landowners by allowing them to receive a share of royalties, usually 12.5 percent for the entire parcel. Thus, a person who owns 10 percent of the land in the parcel would collect one-tenth (1.25 percent) of the 12.5 percent paid to all owners of the mineral rights. The energy industry also claims it protects landowners from not being able to be compensated when natural gas is taken from their land by the horizontal pipes that run beneath that land. Opponents say forced pooling is nothing less than a government seizure of private property rights to benefit private industry.

In Windsor, Colo., several schools could be sitting over the horizontal tubes of the fracking process. "The district has little control over whether the mineral rights owner can drill for oil and gas beneath the district's schools," according to the *Fort Collins Coloradoan*.[316] A study by Western Resource Advocates revealed 26 wells within 1,000 feet of a public school in four counties.[317] The non-profit advocacy group pointed out, "it is illegal in Colorado to idle a vehicle for more than 5 minutes within 1,000 feet of a school— but you can drill for oil and gas, spewing potentially toxic chemicals into the air, as long as you aren't closer than 350 feet."[318] Although the wells must be at least 350 feet away, the underground horizontal drilling could extend as much as a mile. Hundreds of schools in Colorado may be forced to enter into lease agreements with gas drillers to protect the surface ground, Stephanie Watson, assistant superintendent for the Windsor district, told the *Coloradoan*.[319]

Pennsylvania's schools could find themselves in similar positions if forced pooling in the Marcellus Shale becomes law. Pennsylvania's law applies only to the deeper Utica Shale, and does not apply to the Marcellus Shale, which is one reason why the industry is looking to drill into the Utica Shale. A bill (HB 977[320]) to allow forced pooling and to create a state office to regulate its implementation was introduced in the Pennsylvania House of Representatives in 2009. Gov. Ed Rendell gave lukewarm approval, but said he would not sign the bill unless a minimum distance between wells was required and that landowners received "full, fair" compensation.[321] That bill died after three co-sponsors withdrew their support.[322]

A subsequent bill (SB 447[323]), which addressed many of the problems of the previous bill, never left committee. Opposing the bills were 31 environmental and outdoors organizations and Tom Corbett[324] who declared in April 2011 that he would not sign the bill because, "I do not believe in private eminent domain, and forced pooling would be exactly that."[325]

SPLIT ESTATE

The law of Split Estate, dating to the English common law belief that the king owned all the land, allows two different owners to split ownership of surface rights and underground rights. Thus, it is possible that a person who purchases a house, farm, or even a garage, may not have the underground rights, those rights having been held by the original owner or sold to someone else.

According to the Bureau of Land Management (BLM):

"[M]ineral rights are considered the dominant estate, meaning they take precedence over other rights associated with the property, including those associated with owning the surface. However, the mineral owner must show due regard for the

80

interests of the surface estate owner and occupy only those portions of the surface that are reasonably necessary to develop the mineral estate."[326]

The 76-minute documentary, *Split Estate*, focuses upon families of Garfield County, Colo., who, says director/producer Debra Anderson, "struggle against the erosion of their civil liberties, their communities and their health" after the natural gas industry moved into the San Juan Basin. The film, which premiered October 2009 on the cable network, Planet Green, received an Emmy for Outstanding Individual Achievement in a Craft for Research; researchers were Mitchell Marti and Matt Vest; writers were Joe Day, Avery Garnett, and Jean Wendt; Ali MacGraw was the narrator.

"Even reasonably knowledgeable viewers are likely to come away with a heightened understanding of both the politically privileged position of our nation's extraction industries and the role that concerned citizens can play in holding those Industries accountable," wrote Dr. Christopher H. Foreman, professor of public policy at the University of Maryland.[327]

Bill Richardson, governor of New Mexico at the time *Split Estate* was produced, called the film, "an eye-opening examination of the consequences and conflicts that can arise between surface land owners . . . and those who own and extract the energy and mineral rights below. This film is of value to anyone wrestling with rational, sustainable energy policy while preserving the priceless elements of cultural heritage, private enterprise above-ground, and the precious health not only of people but the land itself."[328]

OBSERVATION

There are numerous benefits to the natural gas industry that can invoke eminent domain, forced pooling, and split estate rulings; there may even be some benefits to landowners, even those who object to the drilling yet are compensated for their mineral rights. But, years from now, thousands of landowners who allowed drilling on their property may wonder if the immediate gratification of a few dollars or even sudden wealth was worth the cost of what happened to the health and lifestyle of the people and their environment.

PHOTO: Wendy Lynne Lee

For 12 days in June 2012, homeowners of the Riverdale Mobile Home Village (Jersey Shore, Pa.) and dozens of supporters set up barriers and an extensive media campaign to protest their eviction by a company that planned to build a pumping station and withdraw up to three million gallons of water a day from the Susquehanna River.

CHAPTER 5
Collateral Damage in the Marcellus Shale

PART I
MARCH 2012

There's nothing to suggest that in his 51 years Kevin June should be a leader.

Not from his high school where he dropped out after his freshman year.

Not from his job, where he worked as an auto body technician more than 35 years.

Both of his marriages ended in divorce, but did produce two children, a 31-year-old son and a 28-year-old daughter.

June readily admits that for most of his life, beginning about 14 when he began drinking heavily, he was a drunk. Always beer. Almost always to excess. But, he will quickly tell you how many weeks he has been sober. It's now been 56 weeks, he proudly says.

In October 2008 he was in an auto accident when he swerved to miss a deer and hit an oak tree head on. That's when he learned MRIs showed he had been suffering from degenerative arthritis. Between the accident and the arthritis, he was off work for three months. Then, in May 2009, he was laid off when the company moved.

Unable to work, surviving on disability income that brings him $1,300 a month, just $392.50 above the poverty line, he lives in the Riverdale Mobile Home Village, along the Susquehanna River near Jersey Shore in north-central Pennsylvania. The village has a large green area where families can picnic, relax, and play games, sharing the space with geese and all kinds of animals and natural vegetation.

For most of the six years June lived in the village, he kept to himself—chatting with neighbors now and then, but nothing that would ever suggest he'd be a leader. The last time he led anything was almost two decades earlier when he was president of a 4-wheel club.

On Feb. 18, 2012, the residents found out after reading a story in the Williamsport *Sun-Gazette* their village had been sold and would be demolished. The owner/landlord, Richard A. (Skip) Leonard, later came to each of the 32 families, told them he sold the 12.5 acre park, and they would have two months to leave. It was abrupt. Business-like. "For Sale" signs had been posted at the property almost four years. Kevin June says the residents knew he was planning to sell, "but we all thought it would be to someone who would allow us to stay."

Four days after the residents were ordered to move, certified letters made it official. Leonard had sold the park to Aqua–PVR, a partnership of Aqua America, headquartered in Bryn Mawr, Pa., and the smaller Penn Virginia Resource Partners, with headquarters in Radnor, Pa. Sale price was $550,000. It may have been a bargain—land and industrial parks that have been vacant for years are going for premium sales prices as the natural gas boom in the Marcellus Shale consumes a large part of Pennsylvania and four surrounding states.

Aqua had received permission from the Susquehanna River Basin Commission (SRBC) to withdraw up to three million gallons of water a day from the West Branch of the Susquehanna; the families of the mobile home village would just be in the way. The company intends to initially spend about $20 million to build a pump station and create an 18-mile pipe system to provide fresh water to natural gas companies that use hydraulic fracturing.[329]

Aqua isn't the only company planning to take water in the area. Anadarko E & P and Range Resources–Appalachia have each applied to withdraw up to three million gallons a day from the Susquehanna. While the Delaware River Basin Commission, and the states of New York and New Jersey have imposed moratoriums upon the use of fracking until full health and environmental impacts can be assessed, Pennsylvania and the SRBC have been handing out permits by the gross.

Leonard says he tried to sell the land to someone who would

keep the village, and allow the residents to remain. "I had a legitimate buyer who owned a large mobile home park," he says, but backed out of the deal when he saw the requirements imposed by the Lycoming County Zoning Hearing Board. The Board, says Leonard, required that new tenants of the park, which was in a flood zone, to raise their trailers 12 feet. Doing so would make the trailers inaccessible to anyone with a physical handicap and cost at least $5,000 per trailer to elevate. He says he had letters from the Federal Emergency Management Agency that exempted the trailer park from that requirement, "and the Board said it would check with FEMA to verify it." However, says Leonard, the Board didn't check with FEMA but with the insurance commission, which reaffirmed the Board's contention that trailers had to be elevated 12 feet. "I tried fighting this for over a year," says Leonard, but the Board didn't care. And then Aqua America came along, "and I took their offer."

David Hine, Lycoming County zoning administrator, says FEMA regulations require "any new trailer put onto the property be elevated 12 feet," one and one-half feet above the 100-year flood elevation, "and that applies to anyone in the flood plain." He says Leonard is right about the existing trailers being "grandfathered," but new trailers would have had to meet the elevation requirements. Raising trailers 12 feet, says Hine, "can be done; it just takes some engineering." As far as handicap accessibility, Hine says the Zoning Board isn't concerned because "it doesn't apply to one or two family dwellings."

Daniel Fitzpatrick, local government policy specialist with the Pennsylvania Department of Community and Economic Development, reinforces Hine's statements:

> "Minimum regulations under the National Flood Insurance Program that are contained in a municipality's flood plain management ordinance require that new residential construction in an identified Special Flood Hazard Area be elevated so that the lowest floor is at or above the elevation of the 1% Annual Chance Flood. Placement of manufactured homes qualifies as new construction under the regulations of the program. Failure of a municipality to enforce the regulations in their ordinance could ultimately result in suspension from the program. Placement of manufactured homes in

the flood plain is one of the most dangerous types of development and therefore should receive special attention for compliance."

Some believe there may have been a backroom deal between the zoning commissioners and the natural gas industry, which was becoming ubiquitous in the county. "That's nonsense," says Hine.

Most residents had only a vague knowledge of fracking and what it is doing to the earth. Deb Eck says she "knew nothing" about fracking, "so I 'Googled' it, and up came thousands of hits." She says she "learned real fast" what it was, and what it was doing to the people and the land. Kevin June says the residents of Riverdale "have a lot more knowledge now."

Aqua had originally ordered the residents to leave by May 1, but then extended it to the end of the month. It dangled a $2,500 relocation allowance in its eviction. However, the cost to move a trailer to another park is $6,000–$11,000, plus extra for skirting, sheds, and any handicap-accessible external ramps. But, a third of the trailers can't be moved.

"These are older trailers," says June. His trailer is a 12-by-70, built in 1974, with a tin roof and tin siding ("tin-on-tin"). Most of the trailers aren't sturdy enough to survive a move. But even if all could be moved, there are few places that would take the other families. The parks want the newer trailers, but most parks are full. So, the residents are desperately reading the classified ads for rentals.

Because the natural gas companies are bringing thousands of employees to frack the land, there is a shortage of apartments; most have inflated prices to take advantage of the well-paid roustabouts, drivers, technicians, engineers and other professionals who moved into the area, and spend their money on local businesses eager to improve their own profits. During the past two years, rents have doubled and tripled. "None of us can pay a thousand or more a month," says June. The current mobile home owners paid $200 a month for their lot.

The gas drilling boom that hit Pennsylvania has resulted in an increase in the number of homeless, many of whom were evicted at the end of their leases by landlords who wanted to make apartments available for gas field workers willing to pay higher rents. Many of the homeless, unable to find housing or qualify for jobs in the gas fields, have placed an additional burden upon social service agencies; other homeless, even if employed, are invisible to the people and governments of their communities.

Not long after he was served his own eviction notice, June had a dream. "It was Jesus coming to me, telling me I had to do something," he says. If something goes wrong, the residents have to fix it; Kevin June is the one they call. If he can't fix a problem, he finds someone who can. In this trailer park, as in most communities, there is a lot of talent—"we help each other," says June. "I've had the Holy Spirit running through my veins a long time, but it's running real deep right now," he says.

June is constantly on the move, going from trailer to trailer to try to keep the residents informed, to help families who were abruptly evicted. Whatever their needs, Kevin June tries to provide it, constantly on the phone, running up phone bills he knows he can't afford but does so anyhow because the lives of his neighbors matter.

There's Betty and William Whyne. Betty, 82, began working as a waitress at the age of 13 and now, in retirement, makes artificial Christmas trees. She has a cancerous tumor in the same place where a breast was removed in 1991. William, 72, who was an electrician, carpenter, and plumber before he retired after a heart attack, goes to a dialysis center three times a week, four hours each time. They brought their 12-wide 1965 Fleetwood trailer to the village shortly after the 1972 flood. Like the other residents, they can't afford to move; they can't find adequate housing. "We've looked at everything in about a 30 mile radius," they say. They earn $1,478 a month from retirement, only $252.17 above the federal poverty line.[330] One son is in New Jersey, one is in Texas; the Whynes don't want to leave the area; they shouldn't have to.

There's April and Eric Daniels. She's a stay-at-home mom for their two children; he's a truck driver for Stallion Oil Field Services, delivering water to natural gas companies in Pennsylvania and wastewater into Ohio. Their 14-by-76 foot trailer, which they bought in 2009, is now valued at $13,200; she and her husband were in the process of remodeling it, had already paid $8,000 for improvements in two years, and were about to start building a second bathroom. April Daniels had grown up living in a series of foster houses, "so I know what it's like to move around, but this was my first home, and it's harder for me to leave." Their trailer provides a good home, but can't be moved. "We're pretty much on the verge of just tearing down the trailer and living in a camper," she says. They don't know what will happen. They do know that because of what they see as Aqua's insensitivity, they will lose a lot of money no matter what they do.

Doris Fravel, 82, a widow on a fixed income of $1,326 a month, lived in the village 38 years. She's proud of her 1974 12-wide trailer with the tin roof. "I painted it every year," she says. Eight months earlier, she paid $3,580 for a new air conditioner; she recently paid $3,000 for new insulated skirting. The trailer has new carpeting. Unlike most of the residents, she found housing—a $450 a month efficiency. But it's far smaller than her current home. So she's sold or given away most of what she owns. She may have a buyer for the trailer, but will take Aqua America's offer of $2,500 for it, considerably less than it's worth. "I can't do anything else," she says. "I just can't move my furnishings into the new apartment," she says. Like the other residents, she has family who are helping, but there's only so much help any family can provide. "I never knew I would ever have to leave," she says, but she does want to "see one of those gas men come to my door—and I'd like to punch him in the shoulder."

Not only are there few lots available and apartments are too expensive, but most residents don't qualify for a mortgage; and there are waiting lists for senior citizen and low-income housing. The stories are the same.

No one from Aqua has been in touch with any resident. "If they would just have come out here and talked with us like they were supposed to, there might not have been problems," says April Daniels. But, the company did hire a local real estate agency. The agency claims it has made extraordinary efforts to help the residents find other housing. The residents disagree. April Daniels says "some of the Realtors have gotten real nasty with the people in the park—they just don't understand that we are all in a hardship, so we get mad and frustrated and take it out on them." But there really isn't much anyone can do. The natural gas boom has made affordable housing as obsolete as the anthracite coal that once drove the region's energy economy.

The residents, with limited incomes, have lived good lives; they are good people. They paid their rents and fees on time; they kept up the appearances of their trailers and the land around it. They worked their jobs; they survived. Until they were evicted.

And now it's up to the residents to try to survive. They have become closer; they listen to each other; they hug each other; and the tough men aren't afraid to let others see them cry. "The pain in this park is almost too much at times," says June.

Most of the families see their eviction as a politically-based corporate takeover. June says he went to see State Rep. Garth D. Everett (R-Muncy) "to ask what he could do to help, but his secretary just coldly told me there was nothing that could be done because whoever owns a property can do with it what he wants to do." He never saw the state representative.

During the week Aqua–PVR issued eviction notices, its parent company issued a news release, boasting that its revenue for 2011 was $712 million, a 4.2 percent increase from the year before; its net income was $143.1 million, up 15.4 percent from the previous year.[331] But, for some reason, the company just couldn't find enough money to give the residents a fair moving settlement. "They just expect us to throw our homes into the street and live in tents," says June.

PART II
MAY 2012

The forced eviction had united the residents, but by the middle of May, most reluctantly took the $2,500 relocation allowance and left the village. However, others chose to remain and fight what they saw as an injustice.

Kevin June, becoming adept at how to work with the media, was constantly calling radio and TV stations and newspapers, and working the social media, especially Twitter and Facebook. Alex Lotorto, a grounds technician and volunteer from nearby Lewisburg, Pa., set up a website, SaveRiverdale,[332] with information and photos. The pleas and stories drew assistance from anti-fracking groups Clean Water Action, Earth First, Gas

Drilling Awareness Coalition, and Responsible Drilling Alliance. Soon a nation was able to see micro-documentaries on YouTube, posted by several amateur videojournalists, including Dean Marshall and Cris McConkey, and organizations, among them the Media Mobilizing Project, and a recently-created organization, Saveriverdale.com. Photojournalist Lynn Johnson, who is working on a story about the women of the Marcellus Shale for the *National Geographic*, was in Riverdale to photograph that community's life.

Several readers of the Williamsport *Sun-Gazette* wrote letters to the editor or posted their comments to the on-line site of the newspaper to show their support for the residents of Riverdale.

"I question why the SRBC [Susquehanna River Basin Commission] shows preferential treatment to fretful, whining cries of the gas companies while initiating stress, hardships and inexcusable moving expenses onto families who cannot afford the financial burden to relocate," wrote Weldon C. Cohick Jr., whose ancestors lived in the township for two centuries.[333]

One reader, who identified herself only as Miss Jane, in an online comment noted that her mother went through similar proceedings in Florida. The residents, she said, "were only given 6 weeks notice. [It] got real bad towards the end [when] the company who bought the land did not wait til all the tenants left to start decimating empty trailers. [W]e woke up everyday to terrible destructive noises of units being ripped apart."[334]

Billtown 101 said he would like to know:

> "if the commissioners who approved these plans gave as much consideration to the lives and well-being of those 'Families', as they so quickly accepted a 'Waiver' to overlook the Land Development Ordinance regarding the increased slope of the access road—which will potentially cause storm-water runoff issues? I highly doubt the truck traffic issue played a major role in the selling of your souls. God Bless these families, and I pray that each of you some how will find a greater understanding and Compassion from your local neighbors than you have by the blindness of your local government leaders and the greed of the cold-heart."

However, in the center of the Marcellus Shale boom, most of

the neighbors and readers of the *Sun-Gazette* condemned the residents who had complained about their sudden eviction, and who have refused to leave until they receive what they believe should be fair compensation.

Garder54 called Kevin June "a real scum."

LadyDawg4 called him a "sleazeball."

Proud2bMom called him a "liar and a thief."

The harpies who wrote several hundred posts that appeared online in the *Sun-Gazette* were relentless in their condemnation of the residents. Hiding behind anonymous screen names, the writers, who sound like drunks in a bar fight or callers to an afternoon talk show, could be among the thousands of gas company employees and managers who have moved into the area. They could be those who have leased part of their land to the oil companies. They could also be the business owners who have profited because of selling products to the workers. But most of them condemned the residents.

Linhk48, who posted several dozen times, believes "the new owner's only obligation is to give you notice to vacate. He is under absolutely no obligation to subsidize your move, allow you to live rent free until you move, or hire professionals to help you with relocation. Anything he does is a generosity and SHOULD be appreciated!" Linhk48, like many, called them "rabble-Rousers/troublemakers/trespassers."

Czkb217 thought the police or National Guard could move in, and advised the residents, "SO just pack your stuff and MOVE, you are now breaking the law." It's doubtful any of the commentators know Pennsylvania state law that establishes legal processes that must be met to evict persons from their homes.

CitizenQ, who opposes helping the residents and who posted several times, without evidence claimed "some of the residents have been seen stealing from others."

Linhk48 thought Aqua–PVR should take the residents to court "for leaving the property with trailer shells and trash all over and ask for clean-up costs—and punitive damages after they were so generous."

Several repeatedly questioned where the donations to Riverdale went. Some specifically accused Kevin June of theft and fraud, apparently not having the time or intelligence to learn about the controls and regulations to release money from a bank-held account that is a registered 501(c) charity. "The residents know exactly where the money went and why," says June.

When those writing to the *Sun-Gazette* later learned some of the money was used to buy phone cards, a camera, a weed whacker, and a used $200 riding lawnmower, they increased their assault. Had they taken the time to think or ask questions—something those who type and pound "SEND" often don't do—they would have learned that June used the phone cards to cover many of his expenses from numerous cell phone calls to and from attorneys, the media, and others who had an interest in the problems of the residents. They would have learned that June bought the camera because the lawyers required him to document the appearance of the village and the residents' activities. They would have learned that June bought lawn and gardening equipment because both the previous and new owners had no intention of mowing the lawns or killing the weeds. Cutting grass and eliminating weeds also served to help protect their health; living near the river, with the warm seasons approaching, residents knew there would be increased black fly and mosquito infestations.

Woolrich haughtily wanted to know, "Why on earth would you not have saved money for when you eventually had to move your MOBILE home???" Perhaps, Woolrich, it's because when you have poverty-level income, it's hard to save anything.

No2spanish believed, "[T]his is why people in mobile homes should save their money—instead of spending it on booze or drugs."

Czkb217 thought the residents should have gotten together and bought the park. Since of the families live slightly above the poverty line, they probably didn't have an extra $550,000 plus lawyer fees and closing costs laying around. Nevertheless, Czkb217 believes the residents should "Just man up and put your big boy panties on and MOVE." He objects that his taxes are supporting some of the residents who are using Legal Aid, which receives state and federal funds to assist the impoverished. John Person with the Williamsport office of North Penn Legal Services and Kevin Quisenberry of the Community Justice Project in Pittsburgh assisted the low-income residents of the village; Jonathan Butterfield of the Williamsport law firm of Murphy, Butterfield and Holland assisted *pro bono* for those residents who didn't qualify for legal services.

Justin1 wanted the residents to "Get out of the way of progress already."

Ironically, many of those who supported the fracking industry will soon find out that the new pump station will lead to a loss of well-paying jobs. Aqua–PVR proudly says the new station would be more environmentally friendly since it would remove about 1,000 water hauling trucks per day from the roads.[335]

Many of those who attack the residents and defend corporations probably believe they are good Christians; they attend church regularly and, in one of the more conservative and highly Christian parts of the state, praise God publically.

However, the Rev. Leah Schade, who held an interfaith service at the village, doesn't see them as good Christians. "It is a craven, cowardly way to snipe at people," she says. Those criticizing the residents "are profiting from the way things are or they are so insulated from the pain and suffering the people are undergoing that they are unable to respond with compassion," says Rev. Schade, pastor of the United in Christ Lutheran Church in nearby Lewisburg. "As a Christian," she says, "I make a decision to do what Jesus calls us to do—to minister to those most vulnerable and resist the powers and the principalities that seek their own self perpetuation and their own profit." Rev. Schade, who is completing a Ph.D. in ecological theology, points out, "The church has a long history of offering a prophetic voice to persons who are oppressed and made vulnerable by powerful systems, and who need advocates to speak for and alongside of them in the public arena. The teachings of Jesus would tell us that what is happening to these families isn't right. He would ask, 'Who controls the resources; who does not?' The residents and the surrounding ecosystem are the disempowered ones."

Most of the residents are justifiably afraid to stay and fight Aqua America, and are taking the $2,500 buy-out and trying to find places to live. One by one the residents have stripped their trailers of pipes, fixtures, porches, anything that could be taken and used in a new home, anything that could be sold, anything that other residents might be able to use. Shells of tin and fiber board, some with exposed asbestos, made the place 32 families once called home now looking like a battlefield.

"It's not the trailers that make up the community," says April Daniels, "it's the people who live in it."

PART III
JULY 2012

Aqua America didn't want a fight. It probably didn't even expect the myriad problems that had been dumped upon it. "The guy ["Skip" Leonard] was supposed to have sold us a clean piece of property," Nick DeBenedictis, Aqua America's CEO, told the *Philadelphia Inquirer*.[336] The situation in Riverdale was becoming a public relations and operations disaster. It was David vs. Goliath, and no corporation wants to be seen as Goliath. But Aqua–PVR, under the Aqua America umbrella, wanted to begin construction of the pump station. Some of the Aqua executives had to have felt conflicted, maybe even confused—they had bought the land in good faith, and may not have had knowledge of the consequences; they couldn't understand why 32 families living in trailers would cause such a problem to a huge corporation that just wanted to build a pump station. Besides, they may have reasoned that they weren't even the ones who were fracking the earth; they were merely pulling out water to send to those companies that did the fracking. What Aqua–PVR was doing was relatively clean. Even if it had consequences for the marine life and vegetation, a governmental body had approved its application to withdraw up to three million gallons a day. And now there was a problem that may have seemed greater than what the fracking companies were experiencing.

On Thursday, May 31, the final day residents were legally allowed to remain in Riverdale, seven families remained in what was left of their micro-village. They were joined by 50 anti-fracking activists who showed up to begin a vigil. "We asked the residents about their concerns, explained what we wanted to do, and made sure what we wanted to do would help the residents and not cause them further harm," says Alex Lotorto, one of the protestors.

The next day, the day Aqua said it would begin construction, the protestors blockaded the two entrances to the park. The barriers included old washing machines, tires, cinder blocks, and fiber board, anything the protestors could take out of the trailers that were abandoned. What informally became known as the Occupation of Riverdale began Friday, June 1, 2012.

"It should never have gotten to a blockade," Lotorto says. But it did get to a blockade because of an intractable corporation that was determined to use the land it purchased to put up a pumping station and wanted no delay.

"We are here to fight against the exploitation and abandonment by a society of the economically vulnerable," said Dr. Wendy Lynne Lee, one of the protestors and a leader of Occupy Well Street. "Our purpose was to protect the park's residents from unjust expulsion and to make people aware that this nonviolent protest is against the egregious injustice of these evictions," said Dr. Lee.[337] Seven families remained in the park on the day Aqua had given its deadline.

"When we first heard about the protest," says Eric Daniels, "we [he and his wife, April] weren't very happy because anti-fracking protestors would throw rocks at my truck." However, Daniels quickly says that these protestors weren't involved in any destruction of property. "After sitting and talking with them [the Riverdale protestors] the first day, I realized they were working class people, just like we are. They were there to help us, brought food, shared their lives with us, and cleaned up after residents moved out."

While the residents and protestors were friendly, some of the residents didn't join the protest. Scott Bliler says he "just didn't want to do it." Toby Mainse and his fiancée Traci June, Kevin June's daughter, friends of "Skip" Leonard, refused to be a part of the protest, but had stayed in the village past the June 1 deadline.

Kevin June says the problems in the village for the residents were magnified not by Aqua but by how "Skip" Leonard negotiated the sales contract and the promises he had made to residents that even if he sold the land, none of the residents would have to leave. Kevin June's activism and the subsequent protest would cause tension between Traci June and her father, with Traci not speaking to him at all, says Kevin June.

Another reason some stayed away from the protest, although

sharing their community with the protestors, was because they feared if they joined the protest, it would upset legal negotiations being held with Aqua America for increased compensation; Aqua itself had wanted a separation between the residents and the protestors.

Online comments by readers of the *Sun-Gazette* identified the protestors as "out-of-town activists" or, more specifically, "environmental activists." Bobbie2 called the scene a "liberal zoo . . . a veritable microcosm of the liberal social system." Joe123 called the protestors "unorganized morons," and decided the residents "are on display by 'Fame Seekers', like trick-monkeys in a circus." Proud2bMom, with no facts, something that never stymied any of the others who wrote into the online site, claimed "the residents left that are trying to get out are more or less being held prisoner in their own homes because of the few who feel they need to block the roads." Many, who had never been to the village, called the protestors unwashed hippies who were living off welfare and the government. However, most protestors had jobs, and came to the village on their days off and in the evenings. Some were students from Oberlin College; some were retired.

By the middle of May, there was a subtle change in leadership. Kevin June had been the force to unite the residents and get the media and lawyers, but his communication skills were weak, and residents were becoming confused by what was and was not happening. Deb Eck became the liaison between protestors, residents, police, and on-site company workers. Eck focused upon helping the remaining Riverdale residents before she went to work as a manager of a retail store in Williamsport, worked long shifts, and then came home, usually about 9 or 10 p.m., and worked a few hours. "Some of my days I only got four hours of sleep," she says. June, still doing what he could for the residents, increased his work with the media and the lawyers.

On the sixth day of the protest, "Skip" Leonard, who still owned about 45 acres adjacent to the parcel he had sold, including some camping areas, rode into Riverdale on an orange Kubota tractor, with a front scoop-bucket attached, hit the barricade, an adjacent STOP sign, and almost hit George Vest, one of the protestors. "He hit it, backed up, hit it again and

96

again," says Jackie Wilson, a retired secretary and one of the protestors. On a small mower deck attached to the tractor was a 4-foot x 8-foot plywood sign—"MAKE THE TRESPASSERS CLEAN UP THEIR MESS AND LEAVE!!!—that was later screwed onto a pole owned by Pennsylvania Power & Light.

Wilson says she had asked a State Police sergeant for an incident report to report the damage, but he refused "and was nasty about it." She says "it was a rude awakening," because she had worked 19 years for the Springfield Twp. (Dauphin County) police department, "and never saw anyone treat a person like that."

The protestors and residents understood that Leonard, now retired after working 47 years in construction, was frustrated, upset that he was now seen as a villain who sold out a trailer park. The barriers blocked him from unrestricted access to another part of his property, and he wasn't afraid to make sure others knew he was blocked from a picnic area, even though the protestors removed parts of the barrier any time Leonard had previously showed up to allow him and others to have unobstructed access into and out of the park. "We were trying not to have any conflict with Mr. Leonard and others," says George Daniels, an organic farmer from Lewisburg, Pa., and one of the protestors.

"People were spreading his name around," says Alex Lotorto, "they were mad and upset themselves, they may have been harassing him in town, but the problem still goes back to the Susquehanna River Basin Commission that allowed Aqua America to take water."

But on that one day, the day he ran into a barrier, "Skip" Leonard had had enough. He didn't want to have to go through protestors, who he thought were professional agitators, to reach his picnic area, no matter how easy or hard they had made it for him. Although he no longer owned the land that once was a small trailer park, he was disgusted by its appearance. For 28 years, he had done his best to maintain a park that was pleasant and physically attractive. But now, says Leonard, these protestors "were bringing every kind of garbage they could find and lined that barricade with it; it was disgusting." No police actions were taken against Leonard or the protestors.

97

Most of the residents of Riverdale had respected Leonard. "He was a stand-up guy who ran a safe and clean park," says Eck. "If you were late with the rent, he'd understand, but he'd descend upon you if you left your trash out or didn't mow your lawn." But, says Eck, "he had his days, and how he handled the sale was so unlike everything he was."

The protestors kept modifying the barriers, and a couple of days after the main barrier was hit, Eric Daniels drew up plans of how to improve the barriers, while leaving the access road clear. Daniels, who was still hauling wastewater to Ohio, was questioning the impact of fracking, and by now knew the protestors may have been right in how they were drawing attention to the problems. He willingly ripped off the roof of his trailer to make signs. Residents and protestors worked together to paint the signs, the most powerful one with painted handprints of the children of the village. Another of the signs listed the jobs of the residents.

Matthew West, a sculptor and digital fabricator, helped the residents create many of the signs, banners, and mobile billboards that would go into the barrier. "We worked from 6 in the morning 'til well past 2 in the morning fabricating the riggings and structural support needed to hold such large pieces of material safely in the ground," says West. He says the protestors and residents "worked so hard and quickly because it was important to us to make sure that we responded back to the resistance from the Leonards."

The signs were information; they would be part of the barrier to separate Aqua America from the village, and were placed as billboards adjacent to, but not blocking, the main entrance into the village. The residents and protestors had become closer during the week, but "on that day is the day we [the residents and protestors] bonded, and solidified the sense of community we had all worked to grow," recalls Dr. Lee.

The Riverdale Mobile Home Village was becoming a battle zone. In addition to the barriers, there was now a security tent, a command/media trailer (the one that April and Eric Daniels had owned and abandoned when they left the village), a trailer with a large red cross painted onto it that served as an infirmary (once owned by Summer Gruthoff and Brian Holt), and a trailer (still owned by Kevin June) that served as a community kitchen and a place protesters could get naps, and hot showers. The command trailer and one other trailer (abandoned by Fred Kinley) also served as a place protestors could sleep.

On June 7, the Susquehanna River Basin Commission, having previously approved Aqua America's request to withdraw water from the river, disregarding the voices of residents and protestors who had gone to Harrisburg for the meeting, approved the application to distribute water. The destruction of Riverdale, says Lotorto, "could only have occurred with the approval of the Commission, which didn't seem to have any consideration for the residents."

Tuesday, June 12, 2012, was overcast with intermittent drizzles. Many of the protestors wore light rain ponchos. Only a cold heavy rain could have made this summer day any uglier than it was. Most had known they would eventually be forced to end the protest; most didn't expect it to be this day.

About 10:45 a.m., Deb Eck received a call from one of the lawyers for Riverdale. He told her the police would be at the village about noon to evict the protestors, and to gather all the residents and go back to their homes to avoid being arrested. "But we didn't listen," says Eck. The seven remaining families met at the Bliler's trailer for an emergency meeting.

On site to "neutralize" the protestors were six guards from Huffmaster Security, identified on its web site as a company that is the "leading provider of strike management solutions."[338]

With the confrontation escalating, Jonathan Sidney, a recent graduate of Oberlin College, did his best to reduce the tensions. For four hours, Sidney and others chatted amiably with the security team, asking questions, chatting about this and chatting about that, helping assure there would be no need for confrontation. They took the guards on a walk-through of the village. Like the residents and protestors, the guards were working class individuals, trying to survive in an economy that was only beginning to recover; one had said he was upset about being sent to evict homeowners, having once gone through eviction himself. Others, says Matthew West, were less sympathetic "but were respectful" to protestors and residents; none wanted confrontation.

Shortly after noon, State and local police—about two dozen with handcuffs attached to their belts and an assortment of weapons—and representatives of various companies working in the Marcellus Shale showed up at Riverdale, determined to end the protest and secure the property.

Less confrontational, four persons hired by Aqua America were brought onto the grounds to install a plastic orange fence around the property. A simple job became complicated, however, because the men had begun to place the fence about 20 feet into a right-of-way. "We pointed it out to them," says Jackie Wilson, "and they had to get a surveyor's map to find the property line. The men then put metal fence posts into the ground, while others stayed close, not interfering but not helping. But Wilson had a form of protest many others didn't have. She was 64, and an activist from Occupy Harrisburg, who had camped out 10 of the 12 days of the occupation. She didn't have to say anything to the men. "She would look at the four men with a look of contempt that suggested, 'Does your mother know what you're doing?' " says Dr. Lee.

Eck, as had been her role the past month, was all over the village, chatting with residents, protestors, the security force, and the police, doing her best to keep lines of communication open and not allow what could have been a confrontation. The protestors had no plans to become violent; the police had no reason to know that.

About 1:30 p.m., a State Police trooper asked Eck's help to end the protest. "We want to resolve this peacefully without arrests," she remembers the trooper telling her. But, she also remembers being told that the police had every intention to arrest and jail every one of the protestors. The trooper asked her to talk with the protestors and request they abandon the village.

"We had a verbal agreement with Aqua, and a lot of help and publicity about our situation," she says, "and I didn't want to see anyone arrested." With the police determined to handcuff and remove every protestor, Eck says she saw no way the arrests would continue to help their cause. "The money everyone, including the protestors, raised for us should now be used to help the residents, those who had already left Riverdale and the ones who remained, and not for bail money," she says.

Throughout the 12-day protest, the State Police did not confront the protestors, but showed up now and then to assure there would be no violence or destruction of property. Several residents say the State Police were polite but firm; a couple of protestors believe they were too firm, too brusque. This time, the police told the protestors they could continue their protest, but not the occupation. They had to take down the barriers and if they wished to exercise their First Amendment rights, it would have to be on a berm outside the village.

More confrontational was "Skip" Leonard. On that last day of the protest, Dr. Lee says he told her, "If you don't stop taking pictures, I am going to shove that fucking camera up your fucking ass." Dr. Lee says she responded, "Are you threatening me, Mr. Leonard?" He replied he was. "I was really upset," Leonard later said, "after 10 days, every time you turn around there was someone with a camera in your face. It gets to you." Dr. Lee says although a State Police trooper saw the incident, he walked away. She says the State Police refused to accept a complaint or to issue an incident number. She says an officer from the Jersey Shore Police later told her that she could go to the station to file a complaint.

"We held up our banner, sang, played music, passed out rolls for people to eat, talked with each other—and prepared to be peaceably arrested," recalls Dr. Lee.

About 2 p.m., Deb Eck asked the protestors to end their resistance to avoid being arrested. "Please don't ask us to stand down," Dr. Lee asked of Eck. Dr. Lee says the protestors, about 35, "were prepared to sit down, arm-in-arm in defiance of the police" and continue the protest. But the moment Eck, representing the remaining residents, asked the protestors to "stand down" they were no longer invited guests and now would be arrested as trespassers. "I did not want to give up," says Dr. Lee, "but we were obliged to follow the wishes of the residents."

Dr. Lee remembers, "Some of us cried—perhaps out of the anxiety of the day, perhaps because this beautiful experiment in community was coming to an end, perhaps because the evacuation seemed to mean that Aqua America had won. But nothing could be further from the truth."[339] The residents and the protestors hugged each other, and the 12-day occupation ended.

We had no leverage," says Alex Lotorto, "all it took to end our protest was a call to the State Police [by Aqua] and we were done." Goliath was finally victorious.

Sarah Ross, one of the protestors and a doctoral student in comparative literature/documentary film production at the University of Washington, recalls:

> "Passing through the gate to my car, I thought about how quickly we forged a very real community made up of residents, their families, volunteers, and neighbors. We planted gardens, constructed outdoor stoves, and cleaned up debris leftover from trailers that had been stripped for parts to give families some extra cash for the moving expenses. I thought about how we incorporated roofs and any building materials into the murals, so that even the physical components of the park contributed in promoting the preservation of this community. I thought about how "home" means more than a house—it is comprised of people, it is the land upon which we thrive. Many of us grew up on the Susquehanna River. And, then I thought about how our home had been violated. The Riverdale community invited us into their home, and in twelve short days, it had become ours. I thought about how quickly they [again] tore it down. They threw over our barricades covered in children's handprints, and then they erected a physical dividing line between the residents and all of us."[340]

There was also another dividing line, one not formed from barriers. For years, village residents had formed a loose neighborhood; they were friends and acquaintances. They had picnics and, like residents of all neighborhoods, they sometimes had arguments or just didn't associate with other residents. But, the stress of dealing with Aqua and the forced evacuation led to an increase of rumor and innuendo, with some of the residents verbally challenging others. Everyone was frustrated and tired, their emotions bled raw by dealing with a corporate entity and issues they didn't fully understand. The division also separated those who remained from those who had taken the $2,500 and left. Those who stayed wanted to negotiate a larger moving settlement not only for themselves but also for those who had already moved. But, poor communication, combined with stress and frustration, left those who already moved feeling as if they were abandoned, first by the owner, then by Aqua America, and now by their neighbors.

"I knew this day was coming," said Kevin June, "some of us had to focus on getting out of here as soon as possible, moving on with our lives and trying not to look back."[341]

For the night of the vigil and the succeeding 12 days of the occupation, the protestors were in the village as invited guests. Each day, drivers honked in support as they drove past on Route 220; each day, the protestors drew additional media attention to the problems a small community faced as they struggled to force a corporation to acknowledge that its original terms were unreasonable and needed to be revised.

Late that day, the remaining residents gathered the supplies and possessions hastily left by the protestors and handed them over the recently-constructed fence; a few, protestors and residents, openly wept. The longest around-the-clock continuous protest against fracking and for residents was now over.

At 7 a.m. of what would be the fourteenth day, demolition crews came into the village to take down the barricades and some of the trailers. That's when Alex Lotorto figured out that Aqua America, Range Resources, or the Allan A. Myers Co., contracted to do the demolition, had not conducted an asbestos inspection required by federal law.[342] Lotorto says because he had worked with asbestos, he was aware of the problems and the regulations.

On June 21, nine days after Aqua took full possession, and after repeated calls to the Department of Environmental Protection, the DEP finally sent an inspector to the former village. "He did a cursory inspection," Dr. Lee believed, so she filed a formal complaint. On July 9, the DEP sent Dr. Lee a letter claiming it didn't find any asbestos left in the trailers or on the grounds. "There was no way asbestos wasn't present in some of the trailers," Lotorto says. The asbestos "could have settled into the ground and was then buried," he says. Even if the DEP did find asbestos, the residents were told they would not be allowed to pursue a suit against Aqua America because the residents stripped the trailers and also smoked.

But, there was an even bigger truth that floated not just over Riverdale but the entire state as well. It would be a truth that

103

would continue to establish the cozy relationship between the DEP, a regulatory agency, and the industry it was to regulate. Dr. Lee, who received the official explanation from the DEP, says Andrea Ryder, district supervisor for air quality assurance, told her, "Our job is to educate, not to penalize; we're trying to get people to do it right the next time." Thus, even if there had been open asbestos piles, with the knowledge of the new owners of the former village, there likely would not have been any fines or penalties for violating public health laws.

The second potluck picnic with protestors and former residents, held outside the village on June 17, was the last official event. By then, many of the protestors had maxed out their credit cards to buy food and supplies for themselves and the residents and former residents. But it wasn't a separation; protestors and residents stayed in touch, checking on each other.

Even after Aqua America got what it wanted, it was still giving residents problems. It demanded everyone who left their trailers in the village to give Aqua America clear title. It forbid any of the former protestors from coming into the village to help the residents move. Aqua America also demanded that in addition to excluding the former protestors, only family members could help, and directed Huffmaster to record the license plates and car descriptions of anyone who wanted to help.

On July 7, the day before Deb Eck was to remove her trailer from the land that Aqua America now owned, Huffmaster told her she couldn't bring her mover onto the grounds, "that only my family would be allowed to help." This led to an angry call by the Rev. Leah Schade to Aqua America. The next day, a security guard told her that a professional mover would be allowed to assist. Eck had to call off work, and arrange not only for the move, but for also for others to temporarily care for her five cats and two bearded dragons, all of which she had rescued.

Denise and Scott Bliler asked for an extension to move because Scott was about to have open heart surgery, July 2, to replace an aortic valve. But, Aqua America did not allow that extension. The Blilers had to be out of the village by July 12. No exceptions. Scott Bliler came home from the hospital July 6, but was under doctors' orders not to do any work. Denise's boss had allowed her to take three weeks off, reducing some of her stress. Some of her friends, her son, Robbie, and some of his friends helped. "They were awesome," says Denise, who proudly recalls that every day after work and every day Robbie was off work he and his friends would be at the trailer to do whatever they could to prepare for the move. "It was hectic and stressful," says Denise; that would be a common problem all the residents had because of Aqua America's actions. But, for Denise and Scott Bliler, there was additional stress.

On July 8, with temperatures in the 90s, and with the trailer being towed by a truck from a professional moving company, the Blilers almost managed to get their trailer out of the park safely. "All was going well," says Denise, "until we got to the gate." Because of where the chain link fence, which had replaced the orange plastic fence, was placed, and a gate that was barely wide enough to handle a 14 x 76 foot trailer, one of the biggest at Riverdale, "when the driver tried to turn onto the median to maneuver, an axle broke on the trailer." Within a couple of minutes, while on the median, the hitch broke. The trailer was towed to an automotive garage for repairs.

The last residents to leave the village were Blake and Linda Trimble, July 14, who stayed in a hotel for a month until they could find a trailer they could afford and a park that would accept them.

The residents who had stayed and negotiated with Aqua America received additional financial compensation above the $2,500 that Aqua America had originally promised. However, Aqua America forced the residents to accept a non-disclosure agreement that forbid them from talking about that settlement. The settlement, which none of the residents could talk about, required the residents not to sue Aqua America for any reason, not to be involved in any protest against the company, and not to say anything negative about the company. It also required that those who signed the settlement never speak against Aqua America for any reason at any time or else face legal action. The secret settlement is also believed to have included individual payments of about $12,000, but only for the seven families who had remained in the village after June 1.

Those who left before June 1 received only $2,500. The lawyers and the seven families who stayed at Riverdale after June 1 had vigorously tried to get Aqua America to give full compensation to all families who left before June 1, but the multimillion dollar corporation wouldn't yield.

"I didn't want to sign the agreement," says Deb Eck, who wanted to wait until the report on asbestos was made public, and was upset that the non-disclosure agreement removed her rights of free speech. "That's when I was told," she says, "that if any one of the seven families didn't sign, then no one would get anything. It was all or none." Aqua representatives, said Eck, told us if we didn't sign, they would also go after everyone, including those who took the $2,500 payment and left." Eck believed it was an idle threat, just "a lot of gesturing," but she signed the agreement. "I didn't want the other six families to lose what we have worked for," she says. My pockets were heavier, but I wanted to puke."

Ten families moved into Harvest Moon Trailer Park, near Linden, about five miles east of where the Riverdale village once stood, and not near the river that the residents had so enjoyed. Seven of the families took their trailers; all of them dipped into savings, retirement accounts, or borrowed from their families to be able to afford the move and the $100 a month higher rent at Harvest Moon. Their children would be forced to transfer school districts. Other residents eventually found spaces at other trailer parks, often renting smaller trailers; others are living in the homes of what are now extended families.

Eck laments, "If all of us had stayed and protested, we might have been able to shut that project down or at least get them to allow us to remain." It was wishful thinking; Aqua America had made its decision, and the government planned to enforce it.

As for the place where Riverdale once stood, construction lights have turned the night into daylight, natural vegetation has been cut down, and wildlife has been displaced. But some trailers, now little more than junk but owned by Aqua America, are still on the grounds near the Susquehanna River.

CODA

In May 2012, the Pennsylvania House of Representatives had passed an amendment (HB 1767[343]) to the Manufactured Home Community Rights Act (P.L.1176, No. 261[344]) to benefit residents of mobile home parks who are forced to move. State Rep. Robert Freeman (D-Northampton) says he had originally sponsored the bill because the Barbosa Trailer Park in Bethlehem Twp. was sold to a developer in 2006 and the residents, many of them low-income, were given only 30 days to move.[345] Rep. Freeman had introduced a bill shortly after the sale was announced, but the bill never moved onto the House floor for a vote in five years. He says because the bill was controversial, "We had been in constant negotiation with the industry and advocates since then." The current bill was introduced in June 2011, languished in committee, and may have been brought to the floor not only because of a successful compromise between owners and tenants associations but also because of what had been happening with Riverdale that made the public and legislators more aware of the problems. The House passed the bill, 190–7; the Senate passed it 49–0, and it was signed into law in October 2012.[346] Known as Act 156, the law requires manufactured home community owners to:

- inform residents within 60 days of any decision to close the community;
- inform the Pennsylvania Housing Finance Agency and the home municipality also within 60 days;
- give residents at least six months to leave the community when the closure notice is made
- consider any offer to purchase the community by a resident association representing at least 25 percent of the manufactured home spaces;
- pay relocation expenses of up to $4,000 for single and $6,000 for multi-section manufactured homes;
- pay at least $2,500 or the home's appraised value, whichever is greater, when the homeowner is unable or unwilling to relocate the home; and
- allow tenants to terminate any leases without penalty after receiving the community's closure notice.
- that a judicial process be followed in determining when a mobile home was abandoned.

None of the families from Riverdale Mobile Home Village will be able to take advantage of the legislation. History will record they were just collateral damage.

Aqua America had to destroy a village to develop a pumping station to take as much as three million gallons of water a day from the Susquehanna River and send it to companies that are drilling for natural gas.

PART II:
Health and Environmental Issues

PHOTO: Bob Nilsson

A drilling rig dominates a farm and agricultural land near Troy, Pa.

CHAPTER 6
Public Health Issues

There is no question that horizontal hydraulic fracturing can produce significant quantities of natural gas that can lessen U.S. dependence upon fossil fuels and foreign oil sources. However, both the process itself and human error are "known to produce a variety of physical and chemical hazards that may cause negative health effects," according to a study conducted by a team of public health scientists at the University of Colorado at Denver.[347]

Fluids used in fracking include those that are "potentially hazardous," including volatile organic compounds (VOCs), according to Dr. Christopher Portier, director of the National Center for Environmental Health, a part of the federal Centers for Disease Control (CDC). About 650 of the 750 chemicals used in fracking operations are known carcinogens, according to a report filed with the U.S. House of Representatives in April 2011.[348]

Data from the Texas Cancer Registry in 2009 revealed that six counties in the Dallas–Fort Worth area "had the highest incidence of invasive breast cancer in the state," according to the CDC.[349] The report didn't show a direct link to chemicals used in fracking, but the data did suggest it may have been more than a coincidence between nearby concentrations of nonconventional wells and the high breast cancer rates.

"Associated pollution has reached the stage where it is contaminating essential life support systems," according to the Endocrine Disruption Exchange (TEDX), a national clearinghouse that says it "focuses primarily on the human health and environmental problems caused by low-dose and/or ambient exposure to chemicals that interfere with development and function."[350] The chemicals in the fracking mixture can lead to

111

compromising the neurological, immune, kidney, and cardiovascular systems, according to a team of researchers at TEDX, led by Dr. Theo Colborn who chased down wastewater trucks to get samples she later analyzed to identify chemicals in the fracking mixture. Dr. Colborn concluded that about one-third of all chemicals in the fracking mixture may cause cancer, while almost 90 percent of the toxins in a fracking mixture could cause damage to the skin, eyes, ears, nose, and throat.[351]

Some of the chemicals "are neurological poisons with suspected links to learning deficits in children," while others "are asthma triggers," Dr. Sandra Steingraber told members of the Environmental Conservation and Health committee of the New York State Assembly.[352] Dr. Steingraber, one of the leading authorities and public speakers about the effects of fracking, also pointed out, "Some, especially the radioactive ones, are known to bioaccumulate in milk. Others are reproductive toxicants that can contribute to pregnancy loss."[353]

Tom Bean, a former gas field worker from Williamsport, Pa., says he doesn't know what he and his co-workers were exposed to. He does know it affected his health:

> "You'd constantly have cracked hands, red hands, sore throat, sneezing. All kinds of stuff. Headaches. My biggest one was a nauseating dizzy headache... People were sick all the time ... and then they'd get into trouble for calling off sick. You're in muck and dirt and mud and oil and grease and diesel and chemicals. And you have no idea [what they are] ... It can be anything. You have no idea, but they [Management] don't care... It's like, 'Get the job done.'... You'd be asked to work 15, 18 hour days and you could be so tired that you couldn't keep your eyes open anymore, but it was 'Keep working. Keep working. Keep working.'"[354]

Dr. Amelia Paré, a plastic surgeon, trying to determine the cause of skin lesions for several of her patients living near natural gas wells and wastewater ponds in southwestern Pennsylvania, tested the urine from two families. In one family, she found phenol, hippuric acid, and mandelic acid. In the other family, she found high levels of arsenic.

She says she had asked the Pennsylvania Department of Public Health and the health departments of Allegheny and

Washington counties to test the wastewater and the families' well water, "but they declined." Local anti-fracking activists had provided water buffalos for the families. The lesions decreased in severity and health improved for each of the family members when they didn't use their well water, says Dr. Paré.

"Children are inherently more vulnerable to environmental hazards because their physiology is still developing," the EPA reports.[355] A *Pittsburgh Post-Gazette* story about the health problems of families living near natural gas wells in Washington County noted that nitrate levels were significantly above acceptable health levels.[356] According to the *Post-Gazette*, "Small children can be sickened by drinking water with [high] nitrate levels . . . and infants drinking water or formula with high nitrate levels can die from 'blue baby syndrome.'"[357]

HB 1950, which had preceded Act 13 of 2012, had initially included a provision to provide up to $2 million a year in funding to the Department of Health for "collecting and disseminating information, preparing and conducting health care provider outreach and education and investigating health related complaints and other uses associated with unconventional natural gas production activity."[358] That provision, supported by numerous public health and environmental groups, was deleted in the final bill.

The effect of the refusal of the legislature and governor to adequately fund the Department of Health may have led to a less than adequate response to citizen complaints. The agency not only has failed to respond to citizen concerns, but its responses to the Associated Press have been "at best confusing and at worst misleading."[359]

"No state is requiring enough upfront collection of baseline data and ongoing monitoring to adequately protect local water supplies and public health," according to an investigation published by OMB Watch in July 2012.[360]

More than 250 of New York's leading health professionals, scientists, and environmentalists, along with 66 health and environmental groups, asked New York Gov. Andrew Cuomo to conduct a full health impact assessment (HIA) similar to what Colorado required prior to issuing any permits for natural gas drilling that uses the hydraulic fracturing process. In a 10-page

113

letter with several attachments, sent in October 2011, the signers pointed out:

"In Texas, Wyoming, Louisiana, North Dakota, Pennsylvania, and other states, cases have been documented of worsening health among residents living in proximity to gas wells and infrastructure such as compressor stations and waste pits. Symptoms are wide-ranging, but are typical for exposure to the toxic chemicals and air and water pollutants used in oil and gas development and can often be traced to the onset of such operations."[361]

In calling for the health impact assessment, the group stated that a "comprehensive assessment of health impacts is likely to include information—such as mounting costs for health care and air and water pollution mitigation—that could inform how DEC and other agencies, such as the Department of Health (DOH), evaluate and assess cumulative impacts and how DEC reviews any proposed gas development permit applications."[362]

The Advisory Board of the U.S. Department of Energy concluded in November 2011: "The public deserves assurance that the full economic, environmental and energy security benefits of shale gas development will be realized without sacrificing public health, environmental protection and safety."[363] The report also suggested that the EPA act unilaterally to study the effects of fracking because coordination with state environmental agencies is "not working smoothly."

The American Nurses Association's House of Delegates passed a health care policy statement in June 2012 to use evidence-based information to educate health professionals about the relationships between energy development and health issues. The issue was first raised by the Pennsylvania State Nurses Association (PSNA):

"Human and ecological health risks are directly related to the use of coal-fired power plants, mountaintop removal of coal, offshore and onshore oil and natural gas drilling, and hydraulic fracturing or 'fracking.' Research demonstrates that increased rates of asthma attacks, cardiovascular diseases and lung cancer are all associated with our current reliance on fossil fuels. As our population ages, our vulnerability to these fossil fuel-related exposures will continue to increase.

114

Children are already at higher risk because of their increased susceptibility to respiratory illness."[364]

Dr. Helen Podgainy Bitaxis, a pediatrician in Coraopolis, Pa., who is active in public health issues, says she doesn't want her patients "to be guinea pigs who provide the next generation the statistical proof of health problems as in what happened with those exposed to asbestos or to cigarette smoke."

Denial and Non-Disclosure

The industry's typical first response to a complaint from the public that fracking could contribute to health issues is often to deny any correlation. Industry spokespeople will first claim that the problems existed before fracking was conducted in that area. When additional evidence comes in, they will continue to debate it or offer a settlement. That's because the industry, like most corporations, prefers to settle, often during a process of arbitration, rather than expose itself in a public trial. It's unknown how many families settled claims with the natural gas industry. As a condition of settlement, the companies demand non-disclosure contracts and the sealing of court records,[365] effectively shutting off public comment by the plaintiffs and hiding any public health issues that arose through the discovery portion of the lawsuit. This restriction prevents others from knowing if health problems are confined to one area or may be a problem that affects public health.

However, the public's health may become greater than a company's rights of secrecy. The *Pittsburgh Post-Gazette* and Washington (Pa.) *Observer-Reporter* both filed petitions in August 2011 to intervene in the case of *Stephanie and Chris Hallowich v. Range Resources, et al.*, [C-63-CV-201003954] after Judge Paul Pozonsky agreed with Range Resources and the Hallowichs, who had been reluctant to agree to the company's demand for a non-disclosure clause, to seal the court records.

In its petition to unseal the records, the *Post-Gazette* argued:

"The gas companies' interest in secrecy must yield to the greater social good of disclosing information relevant to public health and safety. Moreover, no Pennsylvania court has ever held that court records may be sealed based on nothing more

115

than the interest in using confidentiality to promote settlements."[366]

In an *amicus curiae* petition, Earthjustice, pointed out:

"When these cases, alleging serious adverse health effects from gas development, are resolved, they are not being resolved in a way that provides more information to the public about the alleged health effects of gas drilling. Instead, the defendant companies are successful at limiting the knowledge of defendants' operations—especially as they relate to public health—gained in litigation to the plaintiffs, who are bound by protective orders and nondisclosure agreements preventing them from sharing such information with the public. Litigation secrecy, like state law limits on disclosure such as Pennsylvania's impact fee law, deprives the public of information that could be used to protect public health. . . .

"Since gas companies use confidentiality so routinely in so many contexts, it is critical to counter this trend by upholding public access to court records in cases involving the health effects of gas development. . . .

"The public interest in accessing the record in this particular case is heightened by the secrecy generally promoted by the natural gas industry. If the industry were more forthcoming generally—if it did not seek exemptions from otherwise applicable federal and state disclosure requirements, did not advocate for and use state laws to limit disclosure of information such as the identity of chemicals used in drilling and fractureing, and did not routinely silence injured parties during litigation or as a condition of settlement—then an order sealing the record here might not be significant. But the calculus changes when an effort to conceal information is part of a pattern and practice limiting dissemination of information on the health impacts of gas development. Against that background, it is all the more important to ensure that health and safety-related information in court records is accessible to the public."[367]

The Superior Court of Pennsylvania agreed, and sent the case back to the lower court in December 2012 for reconsideration.[368]

In June 2012, three families from Wyalusing, Pa., settled with Chesapeake Energy, and refused to accept a non-disclosure agreement. "They wanted the public to know what the settlement was about," said attorney Todd O'Malley.[369] The settlement included paying $1.6 million to Heather and Jared McMicken, Michael and Jonna Phillips, and Scott and Cassie Spencer, which included purchasing their houses and properties.[370] Two years earlier, the families had noticed muddy water in their water wells. Chesapeake denied that its operations caused the problem, claiming excessive methane existed prior to the drilling. A year earlier, in a related case, DEP fined Chesapeake about $1 million for contaminating the water supply of 16 families after expert testimony established there were faulty cement casings that allowed methane to enter the drinking water. A filtration system, provided by Chesapeake to the McMickens, Phillipses, and Spencers did not function properly, according to the three families.

Disclosing Some, but Not All, of the 'Fracking Soup'

Like almost every industry, the natural gas industry, even if it takes public funds and accepts tax benefits from local, state, and federal governments, believes that transparency is something that applies to someone else. It even has federal support to avoid being responsive to the people. The natural gas Industry is exempt from the Emergency Planning and Community Right to Know Act (EPCRA),[371] which allows citizens to learn what chemicals and toxins are used at facilities.

Although most states require disclosure of chemicals used in fracking, except for what the industry claims are "trade secrets," most of the states which allow the fracking process do not require full disclosure of chemicals prior to drilling.[372]

Fourteen of the 29 states in which fracking occurs have no requirements to disclose the chemicals that are in the fracking mixture or that which is brought to the surface by the process itself, according to data compiled by the the Natural Resources Defense Council (NDRC).[373] The chemical disclosure rules that are in place, says the NRDC, may be "woefully inadequate to provide sufficient public health protection—underscoring the need for federal rules that require all oil and gas companies

117

fracking anywhere in the country to fully reveal the chemicals they're using."[374]

In his State of the Union in 2012, President Obama called for full disclosure of all chemicals and compounds used in fracking on federal and Native American lands.[375] The Obama Administration and the Department of the Interior had originally wanted all chemicals and compounds to be disclosed at least 30 days prior to drilling. However, in a concession to the natural gas industry, the proposed rules issued in May 2012 allowed the industry to disclose the chemicals only after drilling at a site had been completed.[376]

"Companies have to realize that they need to be transparent about what they are doing and they need to take the people's concerns seriously," Maria van der Hoeven, director of the International Energy Agency, said. If the companies don't, Van der Hoeven told an audience at Rice University in August 2012, "there's a very real possibility that public opposition to drilling for shale gas will halt the unconventional gas revolution and fracking in its tracks."[377]

With public criticism increasing, the natural gas industry established a website (fracfocus.org[378]) where citizens can learn about fracking issues and most of the chemicals in the fracking "soup." Eight states require gas companies to report all data to the website. However, much of the composition in the fracking toxic soup is not available to the public. Chemicals in wastewater are also not included. A further problem is that the site itself is difficult to navigate, and 29 percent of the chemicals reported by the industry were impossible to track through the Chemical Abstract Service.[379] An investigation by *Bloomberg News* revealed that in the first eight months the website was active (April 11, 2011 through the end of the year), "Energy companies failed to list more than two out of every five fracked wells in [the] eight U.S. states" in the analysis.[380] Those states were Arkansas, Colorado, Louisiana, Montana, Oklahoma, Texas, Utah, and Wyoming, which accounted for 64 percent of all natural gas production in 2010. Since then, Pennsylvania, the 15th largest state in natural gas production in 2007,[381] became the third largest producer of natural gas, after Texas and Colorado.[382]

Because participation in FracFocus is voluntary, corporations decide not just what to report but also when to begin reporting. Thus, some wells that were fracked before April 11, 2011, were not listed on the website. The Bloomberg investigation noted:

"Gaps remain on the website even when wells are disclosed. Companies skip naming certain chemicals when they decide that revealing them would give away what they consider trade secrets. Many of the wells that are listed on FracFocus have at least one or two chemicals marked confidential. Others have far more."[383]

FracFocus was initially funded by oil and gas trade associations and a $1.5 million grant from the U.S. Department of Energy.[384] The website is maintained by two public organizations, the Groundwater Protection Council, an association of state water officials, and the Interstate Oil and Gas Compact Commission.

"FracFocus is just a fig leaf for the industry to be able to say they're doing something in terms of disclosure," said U.S. Rep. Diana DeGette (D-Colo.).[385] Ralph Kisberg, co-founder of The Responsible Drilling Alliance, said he believes the website is little more than "a PR effort to placate people."[386]

Nevertheless, as public pressure for transparency increased and states began demanding more disclosure of the composition of fracking fluids, natural gas companies have been voluntarily submitting more accurate and verifiable data to the website, while still retaining the right not to disclose what they believe are trade secrets.

The natural gas industry has finally released a list of most of the chemicals used in fracking;[387] not released is anything the industry claims are "proprietary" or trade secrets. And, yet, the natural gas industry still believes what it is doing is "safe."

The truth about what's in fracking fluids is revealed by a question from a Canadian politician. Tom Mulcair, leader of Canada's Federal New Democratic Party, asked "If you think that your method of getting to that gas is safe, why won't you reveal the contents of the fracking fluid?"[388] Since the Industry didn't answer the question adequately, Mulcair answered it: "Because that fracking fluid contains known carcinogens and other very dangerous substances."

119

Troy's Suds Depot in Troy, Pa., like many self-wash centers, forbids gas production vehicles because mud and dirt usually contain embedded chemicals and other solid debris that clog the drains and disable wash bays.

PHOTO: Diane Siegmund

CHAPTER 7:
Water Pollution

To frack the earth, energy companies need to siphon massive amounts of water from public rivers and lakes. To do so, they must first get permission from regional or state agencies. In some cases, it isn't difficult to convince public officials to allow energy companies to take what is necessary to drill for oil and natural gas. Even if the water taken from public waterways is clear, by the time it has been used in fracking it is polluted.

The Susquehanna River Basin Commission has routinely approved requests from drillers to remove millions of gallons of water each day from the river, although the commissioners have not requested any health impact statements or undertaken a complete cumulative impact study, says Iris Marie Bloom, an environmental writer and activist.

About 11 million gallons of water a day for fracking operations in Pennsylvania are taken daily from the Susquehanna River, which provides water for about 6.2 million people; about three times the current water withdrawal is anticipated when fracking reaches its peak in the state, says Paul Swartz, Commission executive director.[389] In contrast, the Delaware River Basin Commission (DRBC) has a moratorium on taking water from that river, which provides water to about 17 million people, until health and environmental studies have been completed. About half of all water for New York City is drawn from the river, the largest unfiltered water supply east of the Mississippi River.

"Once the genie's out of the bottle, it could take years if not decades to clean up contamination if we don't get this right," said Collin O'Mara, secretary of the Delaware Department of Environment and Energy, explaining why his state cancelled a meeting of the DRBC, and that his state opposes any attempt to take water from the Delaware River until further studies are done.[390] O'Mara further noted, "[W]e've seen problems in other states where casings have failed. And we keep saying in every way possible that it's much more important to be right than to try to move fast."

Opposing the DRBC and Delaware Gov. Jack Markell, Gov. Tom Corbett, in a press release dated Nov. 18, 2011, explained Pennsylvania's all-out support for natural gas: "Today's delay—driven more by politics than sound science—is a decision to put off the creation of much-needed jobs, to put off securing our energy independence, and to infringe upon the property rights of thousands of Pennsylvanians."[391] Not one statement made by Corbett referred to health and the environment.

In May 2012, Pennsylvania DEP Secretary Michael Krancer, still fuming over the actions to preserve the health of the people of the Delaware River Valley, said Delaware "smells like the tail of a dog."[392] Krancer later said, "I'm not battling with people . . . I'm dialoguing with people."[393] Among the people he didn't "dialogue" with was the Hallowich family.

In November 2007, the Hallowich family—Stephanie, an accountant; Chris, a high school history teacher; Alyson, 5; and Nathan, 3—whose case had led to a right-to-know issue—moved into a newly-built home in Mt. Pleasant Twp., about 25 miles southwest of Pittsburgh. The family had a two-story house constructed upon 10 acres they had previously bought. It was to be their dream home; it soon became their nightmare.

"It's ruined our lives," Chris told the *National Geographic Daily News*. "It's ruined our plans that we had for the kids. It's ruined what we thought was our perfect ten acres," she said.[394] What ruined their lives was a toxic soup from nearby natural gas drilling. Around them, according to the *Daily News*, was "an industrial panorama [of] four natural gas wells, a gas processing plant, a compressor station, buried pipelines, a three-acre plastic-lined holding pond, and a gravel road with heavy truck traffic." For about a year and a half, the family unknowingly drank contaminated well water.[395] Stephanie Hallowich told *Marcellus-Shale*: "We've had a lot of air issues, where it smells really bad, we get burning eyes, burning throats, headaches, ringing ears; we don't know what's coming out."[396]

Range Resources and the Pennsylvania DEP at first denied that the problems the Hollowichs experienced were from fracking. And so the family paid for independent tests. What was coming out, polluting the air and water, were high levels of manganese; from the leaking plastic-lined impoundment dam came acrylonitrile, ethyl benzyl, styrene, tetrachlorethylene, and toluene.[397]

The family, which drank and bathed in polluted water, rented a 1,500 gallon water tank, which they needed to fill every three weeks. The cost exceeded the $300–400 a month they received from Range Resources for mineral rights. Unable to get Range Resources or the DEP to acknowledge what the family now knew, Stephanie took her case to the media. To radio, newspapers, and television; local, regional, and national. But it was her neighbors, the ones who were also exposed to the pollution and health hazards, who complained, not about the companies but about Stephanie. "Our whole community revolves around the church [and] people yell at me when I go there now," she said.[398] Hannah Abelbeck and Elizabeth Berkowitz, of *Faces of Frackland*, explain that the neighbors "see Stephanie drawing attention to her problems as a threat to their interests, since many people expect great royalty checks when their properties are drilled."[399] It is a common problem that divides communities in the Marcellus Shale region.

Unable to sell their house or get compensation for having polluted water, the Hollowichs sued Range Resources, three subcontractors and suppliers, and the DEP.

Matt Pitzarella of Range Resources acknowledged that his company, which now discloses the composition of all chemicals in fracking fluids, may have made "some mistakes—poor communication with a landowner, choosing a bad location for an access road, things like that." But, said Pitzarella, "if we make a mistake we own up to it and make it right."[400] The Hollowichs, said Pitzarella, "are in an absolutely unique situation . . . . Not only will you not see it [lack of communication] in the future, you won't see it now, and you won't see it since then."[401]

Others may dispute that Range Resources improved its communications and significantly reduced human error, or that it "owns up" for every error, but for the Hollowichs—who suffered from pollution and neighbor greed, while a private corporation and a state agency failed to protect their health—the problem finally ended in August 2011 when they agreed to a settlement that, although they reluctantly agreed to allow to be sealed, led to the purchase of their house and property.

The settlement didn't end their problems. Three months after the family agreed to a settlement, they again sued Range Resources, this time for violating the non-disclosure agreement and the gag order imposed by Judge Paul Pozonsky. The Hallowich family claimed the company falsely stated it paid $550,000 in the settlement. According to reporting in the *Pittsburgh Post-Gazette,* the Hallowichs petition "states that Range Resources intentionally and fraudulently filed a Realty Transfer Tax Statement of Value with the state Department of Tax Revenue tax bureau to publicly embarrass Stephanie and Chris Hallowich, inflate the family's tax obligations on the sale of their home, and 'garner a public relations windfall' because the company had paid more than the full market, appraised value of the property."[402] Range Resources denied the allegations.

Tammy Manning, of Franklin Forks, Pa., told Iris Marie Bloom, of *Protecting Our Waters*:

> "All of a sudden our water turned dark grey and then we noticed that it was actually erupting from the well head with a lot of force. You would hear it begin to hiss and then the water would spray out three to four feet in a circle around the well. . . .
>
> "[An official who measured methane levels in our home] told us, as his methane detector was sounding off, that the levels were so high that we should not use the kitchen stove, as it could start a flash fire, and we should leave the bathroom window and door open and fan going during showers, as methane could build up and cause an explosion risk. He also told us the utility companies and fire department would have to be notified of our levels.
>
> "I asked him if we could continue living in our home. He said it was not for him to decide.
>
> "We were concerned that our well might explode and it is very close to the house. So to keep the pressure from building up, we ran the water constantly. Our granddaughter's bedroom is above the kitchen and she began vomiting in the morning when she first woke up. She wasn't running a fever

and after vomiting she was fine. We thought she was just waking up hungry so we left crackers on her night stand.

"By March [2012] our methane levels had nearly doubled. The DEP asked the gas company to vent our well and give us a water buffalo and disconnect our well entirely. Once the well was disconnected, our granddaughter was fine.

"The Friday before our well was vented the DEP tested the free gas coming out of our well and said it was 82% methane coming out. I was quite concerned.

"Also, besides the methane, we had carbon monoxide coming out of faucet. Our water tests also showed very high unnatural levels of some dangerous heavy metals. We bought camp showers for bathing our grandchildren as we were advised that the metals can pose serious health problems and can be absorbed through the skin and inhaled, not just ingested.

"The closest well to us is 4000 feet away. Is a gas lease more important to people than clean water, fresh air, and uncontaminated food?"[403]

In Monroeton, Pa., after a well was drilled near her home in 2009, Jodie Simons said water coming from her well had a "milky grey haze." The water, says Simons, "stinks awfully; it has a scummy, rotten and nasty smell." She told psychologist Diane Siegmund that she and her son soon got rashes with oozing blisters. Simons' daughter, says Siegmund, developed nosebleeds, nausea, and severe headaches. The DEP found higher-than-expected levels of chloride, magnesium, calcium, potassium, and sodium in the water, says Siegmund, but told the family the water was safe to drink.

The problems the Hallowichs, Tammy Manning, and the Simons family faced aren't isolated cases, but are only one part of the story of water pollution from fracking operations. There are thousands of other problems. Residents living near gas wells "have filed over 1,000 complaints of tainted water, severe illnesses, livestock deaths, and fish kills," according to the Environmental News Service.[404] A *ProPublica* investigation in 2009 revealed methane contamination was widespread in drinking water in areas around fracking operations in Texas, Colorado, Wyoming, and Pennsylvania.[405]

The American Academy of Pediatrics recommended "families with private drinking water wells in NGE/HF [natural gas extraction/hydraulic fracturing] areas should consider testing the wells before drilling begins and on a regular basis thereafter for chloride, sodium, barium, strontium, and VOCs."[406]

Methane in the Nation's Water Supply

Methane, a greenhouse gas, is the primary chemical in natural gas. As a compressed natural gas, it can be used for heating, cooking, and as a fuel for transportation.

However, there are several problems with methane escaping into the air and water. Methane has a global-warming potential (GWP) of 72 over a 20 year period, according to the Intergovernmental Panel on Climate Change.[407] This means that over 20 years, methane will trap 72 times as much heat as the same amount of carbon dioxide. This is critical because environmental scientists believe if the current and projected levels of methane enter the system within the next 20 years, it could raise the earth's overall temperature by 2 degrees; this would be a "point of no return."

Methane itself isn't toxic, but is highly flammable; in large concentrations in water it will replace oxygen and can cause asphyxiation. Numerous explosions of freshwater wells near natural gas rigs attest to the problem that develops when fracking is used to bring methane from oxygen-deprived shale to the surface.

"Today's methods make gas drilling a filthy business. You know it's bad when nearby residents can light the water coming out of their tap on fire," says Larry Schweiger, president of the National Wildlife Federation. The problem is magnified because one-fifth of all Pennsylvania's residences,[408] second highest number in the nation,[409] draw their drinking water from shallow wells. There are no Pennsylvania regulations about construction or maintenance of private drinking wells, although well drillers are required to submit well completion data to the Department of Conservation and Natural Resources. Pennsylvania's water wells also tend to have higher concentrations of numerous minerals and chemicals than the standard suggested by the Department of Environmental Protection.

A natural gas explosion damaged a house in Bainbridge

Twp., Ohio, Dec. 15, 2007. There were no injuries. "Early in the investigation, responders recognized that natural gas was entering homes via water wells," leading to the evacuation of 19 homes,[410] according to the Department of Natural Resources (DNR). During the subsequent week, Dominion East Ohio, the contractor, said it "disconnected 26 water wells, purged gas from domestic plumbing/heating systems, installed vents on six water wells, plugged abandoned in-house water wells, plumbed 26 houses to temporary water supplies, provided 49 in-house ethane monitoring systems for homeowner installation, and [provided] bottled drinking water to 48 residences."[411] The DNR determined the primary contributing factors to the explosion were inadequate cementing on the casing, a decision to proceed with the fracking process "without addressing the issue of minimal cement behind the production casing," and a 31 day period following fracking that "confined the deep, high pressure gas . . . within [a] restricted space."[412]

In December 2010, the EPA ordered Range Production to protect families and their water supply after determining there was natural gas in drinking wells that serve two homes in Parker County, Texas, and that explosions were "imminent." Al Armendariz, EPA regional administrator, says the EPA acted when Texas regulators "acknowledge[d] that there is natural gas in the drinking water wells," but delayed action. Range Production had used fracking on a well near the two homes. While Range Production denied it caused the problem [See: *Hearing Before Texas Railroad Commission*[413]], Armendariz was blunt in his assessment. "We know they've polluted the aquifer," he said.[414] In March 2012, the EPA withdrew its Imminent and Substantial Endangerment Administrative Order, thus ending all EPA actions. [See: *U.S. v. Range Production.*[415]]

In late 2010, equipment failure may have led to high levels of chemicals in the well water of at least a dozen families in Conoquenessing Twp. in Butler County, Pa. Township officials and Rex Energy, although acknowledging that two drilling wells had problems with the casings, claimed there were pollutants in the drinking water before Rex moved into the area. John Fair disagrees. "Everybody had good water a year ago," Fair told environmental writer and activist Iris Marie Bloom in

February 2012. Bloom says residents told her the color of water changed (to red, orange, and gray) after Rex began drilling. Among chemicals detected in the well water, in addition to methane, were ammonia, arsenic, chloromethane, iron, manganese, t-butyl alcohol, and toluene.[416] Not acknowledging that its actions could have caused the pollution, Rex did provide fresh water to the residents but then stopped doing so on Feb. 29, 2012, after the Pennsylvania DEP said the well water was safe. The residents vigorously disagreed and staged protests against Rex.[417] Environmental activists and other residents trucked in portable water jugs to help the affected families. Joseph P. McMurry of the *Marcellus Outreach Butler* blog (MOB) stated that residents' "lives have been severely disrupted and their health has been severely impacted. To unceremoniously 'close the book' on investigations into their troubles when so many indicators point to the culpability of the gas industry for the disruption of their lives is unconscionable."[418]

In May 2011, the Pennsylvania DEP fined Chesapeake Energy $900,000, the largest amount in the state's history, for allowing methane gas to pollute the drinking water of 16 families in Bradford County the previous year.[419] The DEP noted there may have been methane emissions from as many as six wells in five towns. The DEP also fined Chesapeake $188,000 for a fire at a well in Washington County that injured three workers.

In May 2012, high levels of methane seeped into streams and private water wells in homes in rural Leroy Twp. in Bradford County, Pa., from a well owned by Chesapeake Energy. Human and mechanical error contributed to the escape of methane gas, according to the DEP, which responded after a complaint by the Clean Air Council. The CAC had commissioned a study that showed twice the normal air-borne methane levels.[420] Dr. Bryce Payne, an environmental scientist who monitored air and water pollution in a two square mile area, explained to the Clean Air Council:

> "Two methane plumes were detected. One larger plume substantially increased in size over a few hours, which suggests large amounts of methane were being emitted into the air. A smaller plume was also detected about 2 miles west of the larger one. The data and observations suggest natural

gas has spread through an extensive underground area beyond where the plumes were found."[421]

However, DEP Secretary Michael Krancer, in a letter to the CAC in July, claimed, "The situation is, and at all times was, under control by DEP. Indeed, at this point in time the situation is for the most part over."[422] But the situation was not "for the most part over."

Field research by Gas Safety two weeks after Krancer's letter revealed, "methane concentrations as high as 94 percent just below the soil surface; an airborne methane plume covering about 1.6 square miles; and bubbling in Towanda Creek."[423] Dr. Payne, co-author of the Gas Safety report, told the AP, "[It] is clear that at this point the event and the damage to groundwater and the domestic wells it supplies is certainly not over, and there is no foreseeable end in sight."[424]

Dr. Payne told the Scranton *Times-Tribune* the company and DEP "are not adequately addressing that issue [of impact beyond 2,500 feet]. And they are not willing to provide any information to indicate why it is that they are concluding that everything is getting better and better."[425] Krancer's response, provided by a deputy press secretary as obdurate as Krancer himself, was "As Secretary Krancer pointed out in his letter, a letter that is still entirely accurate and that we stand by, we have an active investigation under way to monitor the situation as it unfolds."[426]

The Epicenter of Pennsylvania's Gas Production

Most of the residents of Dimock Twp. Pa., a rural area with about 1,400 residents in Susquehanna County, are farmers and blue collar workers struggling in a depressed economy. The largest permanent employers in the county are the schools, government, and a hospital. About 13 percent of the population was below the poverty line, according to the 2000 census. And then Cabot Oil & Gas showed up, offering thousands of dollars for leases and royalties for mineral rights. The natural gas boom helped revitalize Dimock and brought a temporary prosperity to many of its landowners, who quickly became believers in natural gas exploration. But all was not peaceful.

Fifteen residents sued Cabot Oil & Gas, in November 2009, claiming the company contaminated their drinking water.[427] The cause was failure of well casings. Tests conducted by the Pennsylvania DEP during the last years of the Ed Rendell administration had revealed there was higher than expected methane gas in 18 wells that provided drinking water to 13 homes near the gas wells. The build-up of methane gas had also led to well explosions and DEP warnings to citizens to keep their windows open.[428] Among the provisions of a consent order, the state required Cabot to provide fresh water to those families whose water had been affected by the excess methane gas. Cabot denied its fracking operation was responsible for the elevated levels.[429] On Nov. 30, 2011, after the DEP, now under the Tom Corbett administration, declared the water to be safe to drink, Cabot stopped delivering water.

And then something strange happened. The town of Binghamton, N.Y., about 35 miles north, said it would provide a tanker of fresh water to the residents who were affected by the drilling. However, the supervisors of Dimock Twp., supported by most of the 140 residents who attended the meeting, most of them with some economic ties to the natural gas industry, refused the offer. When Binghamton mayor Matthew T. Ryan asked "Why not let people help?" he was rebuffed by one of the township's three supervisors who snapped, "Why should we haul them water? They got themselves into this. You keep your nose in Binghamton."[430]

In January 2012, after declaring the water "contains levels of contaminants that pose a health concern"—including arsenic barium, manganese, and other toxins—the EPA decided it would bring water to residents in Dimock. Cabot's response was that the EPA was wasting taxpayer money in its investigation of Cabot's environmental and health practices.[431] The response by Pennsylvania's DEP was almost as inflammatory as the water in the taps. Michael Krancer, the DEP head, disagreed with the EPA findings, and called the agency's knowledge of fracking to be "rudimentary."[432]

In his second meeting with a Congressional panel, this time on May 31, 2012, Krancer increased his insistence that the federal government should defer to the states. The federal government should not have a pre-emptive role, said Krancer, who argued, "The question is a fundamental one: Are you in a better place in Washington to tell us what to do?"[433] This prompted Rep. Gerry Connolly (D-Va.) to fire back at Krancer's "states' rights" arguments: "Those are the same kinds of arguments that were used for generations. If we were talking 40 to 50 years ago about Jim Crow laws in the South and civil rights, we wouldn't have heard testimony at this table." Krancer later declared the EPA was "rogue and out-of-control."[434]

In mid-March, following preliminary tests on several of the wells serving Dimock residents, the EPA claimed the water "did not show levels of contamination that could present a health concern."[435] However, it acknowledged arsenic, some metals, and potentially explosive methane gas remained in the water. A *ProPublica* investigation revealed that four of the five water samples it obtained showed methane levels exceeding Pennsylvania standards.[436]

"We are deeply troubled by Region 3's rush to judge the science before testing is even complete, and by their apparent disregard for established standards of drinking water safety," said Claire Sandberg, executive director of Water Defense.[437] She questioned why EPA Region 3's handling of the Dimock case differed from how other EPA regional offices handled similar cases in Texas and Wyoming when it didn't release the information until all testing was completed. Dr. Ronald Bishop, professor of chemistry at the State University of New York at Oneonta, told *ProPublica*, "Any suggestion that water from these wells is safe for domestic use would be preliminary or inappropriate."[438]

In July 2012, 32 families and Cabot reached a verbal agreement for financial settlement. In a public statement, Victoria Switzer, who had been outspoken in her condemnation of how Cabot dismissed concerns about health and environment, issued a brief public statement that said her family "was relieved to put this behind us and hopeful that we will be able to live out our lives in the home we have invested so much of our time and resources in."[439] In what may have been seen as a public relations statement, she now suggested she would "advise anyone living in a gas field with concerns or disputes involving a gas company to try to work with them." However, four families in the suit chose not to "work with" Cabot. Ray

Kemble said he didn't settle because terms of the agreement would restrict what he could say publically.[440] Kemble buys fresh water, afraid to use the local water that had turned brown; his property has anti-Cabot and anti-fracking signs.[441]

The same month that lawyers for Cabot and the residents announced their tentative agreement, the EPA declared that water in Dimock was safe to drink, and allowed Cabot to stop delivering fresh water to four families. The following month, the Pennsylvania DEP allowed Cabot to resume fracking in the seven wells that were already drilled, after almost a two and one-half year suspension.[442] However, the long-term effects from contamination from that water for several families will always be questioned. Of all companies drilling in Pennsylvania, Cabot leads the list, as of the beginning of 2013, with the amount of fines and penalties ($3.2 million), and is second (to Chesapeake) with the number of violations (476).[443]

The problems for Dimock residents may not have ended just because one state and one federal agency said the water was safe to drink. The Agency for Toxic Substances and Disease Registry (ATSDR), an agency of the U.S. Department of Health and Human Services, based upon extensive analysis of data of metals in the water—the DEP was primarily interested in methane contamination—concluded at the end of December 2011 that not only are there "important data gaps for evaluating water quality in private wells," but that:

> "there is a possible chronic public health threat based on prolonged use of the water from at least some of these wells— assuming future exposure to these contaminants at these concentrations is not reduced. Based on the potential quality control issues, a potential health threat for the remaining wells cannot be disregarded. Additional characterization of the ground-water quality and a thorough review of any changes in concentration over time are indicated."[444]

It recommended there be no further use of the well water "sampled to date at this site," and that "distribution of alternative residential water supplies should be considered until potential exposures are further understood and mitigated as needed."[445]

Accidents and Intentional Pollution

About 11:45 p.m., April 19, 2011, a blowout at a Chesapeake Energy well (Atgas 2H) about 13 miles west of Towanda in Bradford County, Pa., spilled several thousand gallons of frack fluid and other emissions into Towanda Creek, which flows into the Susquehanna River.[446] Chesapeake's Brian Grove, two days after the blowout, told the Scranton *Times–Tribune*, "Initial testing from Towanda Creek indicates little, if any, significant effect to local waterways as a result of an apparent surface equipment failure [and] fluids from the well are fully contained."[447] However, containment would not be for another three days; Grove's statement of how thousands of gallons of toxic chemicals spilled onto agricultural fields and into a creek could be negligible strains the level of reason.

Nevertheless, there was another problem with the gas well's blowout that magnified the problem. It took about 13 hours for an emergency response team to arrive on site; the team had to fly from Texas into Pennsylvania.[448] John Hanger, former DEP secretary, told *ProPublica* that the state had a contract with a private company, CUDD Well Control, to provide emergency response to any place in the state in less than five hours.[449] CUDD had an emergency operation based in Bradford County, site of the well blowout. A CUDD executive, according to *ProPublica*, had offered its assistance, but the DEP didn't request its help; Chesapeake also rejected CUDD's assistance, and called in from Texas another company, Boots and Coots, owned by Halliburton. The DEP and Secretary Michael Krancer "didn't respond to calls and emails from *ProPublica*."[450]

Rory Sweeney of Chesapeake later told *ProPublica* that an in-house control specialist was at the site within 30 minutes and three more arrived within eight hours, and reduced the flow by 70 percent before the Texas team arrived. However, the DEP, in issuing a violation, asked "why Chesapeake took 12 hours to have a well control service company at the site when there are other well control service companies located closer to the Atlas 2H Well."[451]

The Scranton *Times-Tribune*, citing official DEP records, reported Chesapeake "has been issued 30 notices of violations from the DEP for its operations in the state this year. The company has been cited 284 times for violations since the start of 2008 and has been subject to 58 enforcement actions by environmental regulators."[452]

About two weeks after the spill, Maryland Attorney General Douglas F. Gansler, with support from Gov. Martin O'Malley and Maryland's Department of the Environment, notified Chesapeake it was filing suit against the energy company[453] for violating the Resource Conservation and Recovery Act (RCRA) and the Clean Water Act (CWA). The Susquehanna, according to the suit, provides about 45 percent of the freshwater of the Chesapeake Bay, giving Maryland legal standing to sue. Although the contaminated water was eventually contained and did not enter the Chesapeake Bay, Chesapeake Energy agreed in June 2012 to contribute $500,000 to the Susquehanna River Basin Commission to assist in water quality monitoring, and to change some of its practices that impact water quality and the environment.[454]

Four months later, following a two-year vigorous investigation by the EPA, Chesapeake Appalachia, a division of Chesapeake Energy, paid a $600,000 fine for water pollution in West Virginia.[455] The U.S. District Court accepted Chesapeake's guilty plea on three criminal violations of the Clean Water Act. Court documents state that the corporation hired contractors who polluted the wetlands, and dumped 60 tons of crushed stone and gravel in December 2008 that buried a natural waterfall. Chesapeake's reason for the violation was that it needed a path for its trucks carrying water and other supplies to a natural gas fracking site. The fine, one of the largest handed down in a criminal case prosecuted by the EPA and Department of Justice, represented 0.00571 percent of the corporation's $10.5 billion profit in 2011.

In January 2012, equipment failure at a drill site in Susquehanna County led to a spill of several thousand gallons of fluid for almost a half-hour, causing "potential pollution," according to the DEP. In its citation to Carizzo Oil and Gas, the DEP "strongly" recommended that the company cease drilling at all 67 wells "until the cause of this problem and a solution are identified."[456]

In some cases, it's relatively easy to find a cause and solution to water pollution. Between 2003 and 2009, Robert Allan Shipman and his company, Allan's Waste Water Service, illegally dumped several million gallons of wastewater, sludge, and grease onto the ground and streams in six southwestern Pennsylvania counties. "He was pouring the stuff in any hole he could find," said Nils Frederiksen, press secretary for the state's Office of the Attorney General.[457] "God only knows where he put it all. What we found was incredible. But who knows where else . . . all the other places he dumped it," said Greene County Sheriff Richard Ketchum.[458]

According to the Grand Jury presentment, "This activity would typically occur after dark or during heavy rain so that no one would observe the illegal discharge."[459] The state charged Shipman with 98 criminal counts, including criminal conspiracy, theft by deception, receiving stolen property, forgery, tampering with public records; the state charged Shipman's company with 77 criminal counts. Information in the Grand Jury presentment charged:

"The drivers stated that every action they took while working for Allan's Waste Water was done at Shipman's direction. Some drivers testified they believed they were fired for their unwillingness to engage in illegal activities. Other drivers voluntarily quit for the same reason. Most drivers complied with Shipman's requests because they did not want to lose their jobs."[460]

In February 2012, Shipman plea bargained to being charged with only 13 criminal charges, and accepted 13 charges against his company. Judge Farley Toothman sentenced Shipman only to seven years probation, fines, restitution, and a requirement to work five hours a week for seven years with a water conservation group. Deputy Attorney General Amy Carnicella called the sentence too lenient. In December, the Office of the Attorney General formally appealed to Superior Court, claiming the sentence "did not fit the crime" and should have carried at least 16 months imprisonment, according to state sentencing guidelines.[461]

While it's understandable that some employees were willing to overlook illegal activities in order to cling to their jobs in a depressed economy, there were others in the community who knew about the pollution, yet did nothing to stop it. Aaron Skirboll of *AlterNet* noted, "When the arrest finally came in March 2011, area residents were elated that the man behind the worst kept secret in Greene County was going to pay for his crimes."[462] These unindicted co-conspirators had little to be elated about; they were the ones who knew, yet did nothing.

The Pennsylvania Fish and Boat Commission several times asked the DEP to place the Susquehanna River on the "impaired waterways" list, which would mean it would receive a priority to be cleaned up. The Commission was specifically concerned about what appeared to be black skin lesions and diseases affecting small mouth bass. However, Michael Krancer refused the request. Krancer told the Commission, in a letter made public by the *Sunbury Daily Item*, "the lesions and sores on the fish . . . are a complex problem and the reasons are not fully understood."[463] Dr. William Yingling, a physician and fisherman, disagreed:

> "The evidence points strongly to the fact that the problems . . . are being caused by chemical endocrine disruption pollution in the watershed. If these black skin lesions on the smallmouth bass are spindle cell tumors or show cellular changes of melanoma, this presents serious questions about human cancer risks from the chemical pollution in the watershed. We have been told [by DEP] that the black spots will not harm us and the fish are safe to eat. That may be correct. But if the spots represent the influence of chemicals in the water and they show evidence of malignant change in the fish, what would be the effect on human health?"[464]

Dr. Yingling suggested that histopathologists and scientists from the U.S. Geological Survey and the National Institute of Environmental Health Sciences analyze the problem. Krancer told the *Daily Item* the DEP added water quality gauges that "will continue to operate and staff will assist when they can,"[465] but refused to even speculate that water pollution caused by fracking could have contributed to the health problems of fish. Nevertheless, data strongly suggests the problem was not present before fracking operations began on a large scale.

Stuart Gansell, former director of the Bureau of Watershed Management, on behalf of 22 retired DEP scientists, engineers, and senior administrators, wrote a letter to Krancer, telling him they didn't understand why the DEP refused to declare the river was polluted. In opposition to Krancer's belief, they wrote, "It is not necessary to know the reason for the impairment. Listing would focus attention and funding on the issue. This, in turn, will help to resolve the problem."[466] Krancer's response to DEP professionals was as dismissive as his rejection of the Fish and Boat Commission's request. Declaring 90 miles of the Susquehanna as polluted would be nothing more than a "publicity stunt," Krancer replied, and suggested that many of the concerned former DEP staff didn't have the professional quailfications to fully understand reasons why he refused to declare the Susquehanna was polluted.

A week after Gansell's letter was sent, the Fish and Boat Commission filed a formal request to the DEP to reconsider its position to not declare portions of the Susquehanna to be polluted and to reclassify a 90 mile portion as "a high-priority impaired and threatened river."[467] Krancer rejected that request.

Mario Salazar, a former EPA engineer, told *ProPublica*, "In 10 to 100 years we are going to find out that most of our groundwater is polluted. A lot of people are going to get sick, and a lot of people may die."[468]

Academic Research

SUPPORTING THE ENERGY INDUSTRY

The natural gas industry needs to "seek out academic studies and champion with universities—because that again provides tremendous credibility to the overall process," said S. Dennis Holbrook, executive vice president and chief legal officer of Norse Energy and a member of the board of directors of the Independent Oil and Gas Association of New York (IOGA).[469] Holbrook pointed out, at a conference in Houston, Oct. 31–Nov. 1, 2011, that IOGA has "done a variety of . . . activities where we've gotten the academics to sponsor programs and bring in people for public sessions to educate them on a variety of different topics."[470]

One of the ways IOGA helped direct academic research is by its connection to SUNY's Shale Resources and Society Institute (SRSI), which sponsored lectures, workshops, and professional papers. Among those papers was "Environmental Impacts During Shale Gas Drilling: Causes, Impacts and Remedies."[471] According to Steve Horn, who wrote an extensive analysis of the SUNY/Buffalo ties to the Marcellus Shale Industry, "Calling the final product a 'study' is a generous way of putting it. . . . [A]ll four co-authors had ties to the oil and gas industry, as did four of five of its peer reviewers. The study didn't contain any acknowledgement of these ties."[472]

The Public Accountability Project analyzed the SUNY/Buffalo study and "identified a number of problems that undermine its conclusion." Among the problems were:

> ". . . data in the report shows that the likelihood of major environmental events has actually gone up, contradicting the report's central claim; entire passages were lifted from an explicitly pro-fracking Manhattan Institute report; and report's authors and reviewers have extensive ties to the natural gas industry. . . .
>
> "[T]he serious flaws in the report, industry-friendly spin, strong industry ties, and fundraising plans raise serious questions about the Shale Resources and Society Institute's independence and the University at Buffalo's decision to lend its independent, academic authority to the Institute's work."[473]

In November 2012, six months after the Institute was created, SUNY/Buffalo closed it, after faculty and environmental groups questioned the Institute's independence and academic integrity.[474]

Research by a team of Penn State scientists concluded:

> "Dissolved methane did not increase at fracked sites and was not correlated to the distance to the nearest Marcellus well site. . . .
>
> "Results of the water quality parameters measured in this study do not indicate any obvious influence from fracking in gas wells on nearby private water well quality. Data from a limited number of wells also did not suggest a negative influence of fracking on dissolved methane in water wells."[475]

However, the study's authors also pointed out:

"[I]it is important to note that this study largely focused on potential changes within a relatively short time period (usually less than six months) after fracking occurred, given the timeline of the project's funding. More detailed, longer-term studies are needed to provide a more thorough examination of potential problems related to fracking, and to investigate changes that might occur over longer time periods."[476]

The Penn State study was funded by the Center for Rural Pennsylvania, a legislative agency of the Pennsylvania General Assembly.

Like the Penn State study, research conducted by Drs. Charles G. Groat and Thomas W. Grimshaw and a team from the Energy Institute at the University of Texas, placed the primary problem of methane in well water with the construction problems in both natural gas wells and drinking water wells, rather than the process itself. The authors stated in February 2012, "[T]here is at present little or no evidence of groundwater contamination from hydraulic fracturing of shales at normal depths."[477] The presence of methane and toxins in well water, said the research team, probably pre-dated natural gas fracking and that a major problem is well construction— Pennsylvania and Alaska are the only two states that do not have regulations for fresh water well construction:

"[M]any of the water quality changes observed in water wells in a similar time frame as shale gas operations may be due to mobilization of constituents that were already present in the wells by energy (vibrations and pressure pulses) put into the ground during drilling and other operations rather than by hydraulic fracturing fluids or leakage from the well casing. As the vibrations and pressure changes disturb the wells, accumulated particles of iron and manganese oxides, as well as other materials on the casing wall and well bottom, may become agitated into suspension causing changes in color (red, orange or gold), increasing turbidity, and release of odors. . . .

"The greatest potential for impacts from a shale gas well appears to be from failure of the well integrity, with leakage

139

into an aquifer of fluids that flow upward in the annulus between the casing and the borehole . . . . In general, a loss of well integrity and associated leakage has been the greatest concern for natural gas–leading to home explosions as described in a subsequent section. . . .

"[I]n most cases, [explosions of fresh water wells are] the result of naturally-occurring methane migration into aquifers and wells before shale gas development began."[478]

Essentially, Dr. Groat's study supported the industry's claims that fracking doesn't cause health and pollution problems.

However, the Public Accountability Initiative revealed in July 2012 that research by Dr. Groat may have been compromised by a conflict of interest.[479] The Initiative, a non-profit group, disclosed that Dr. Groat is a member of the board of Plains Exploration and Production Co., which conducts fracking operations. He received an annual fee for being a member of the Board; in 2011, it was $58,500, according to *Bloomberg News*.[480] Since November 2007, when he became a member of the Board, Groat received about $1.6 million in stock from the company. The Initiative noted that the research by Dr. Groat and his team was distinguished by "bold, definitive, industry-friendly claims highlighted in the press release but not supported by the underlying report; evidence of poor scholarship and industry bias; and dubious and inaccurate claims of peer review" that had led the media to report there was no relationship between fracking and health and pollution problems. In response, Dr. Groat said his role "was to organize [the study], coordinate the activities and report their conclusions."[481] He claimed he did not "alter their conclusions" and his presence on the Pioneer board had "no bearing on the results of the study."[482]

An independent investigation initiated by the University of Texas, and released in December 2012, found "failures and inadequacies in several procedural areas,"[483] and that the study "fell short of contemporary standards for scientific work."[484]

Dr. Groat retired from the university the month before the report was released, and became director of the Water Institute of the Gulf. Dr. Raymond Orbach, the Energy Institute's director who had no participation in the project, resigned; he was not under investigation and kept his tenured status on the faculty.[485]

140

Several other research studies conducted at American universities and funded by either gas/oil companies or their front organizations allowed Barry Russell, president of the Independent Petroleum Association of America, to claim, "no evidence directly connects injection of fracking fluid into shale with aquifer contamination." Elizabeth Ames Jones, former chair of the Railroad Commission of Texas claims it is "geologically impossible for fracturing fluid or natural gas to migrate upward through thousands of feet of rock."[486] Fracking "has never been found to contaminate a water well," says Christine Cronkright, communications director for the Pennsylvania Department of Health.

Independent research studies and numerous incidents of water contamination prove otherwise.

INDEPENDENT RESEARCH REVEALS THE TRUTH

"To naively believe that you can put 4.5 millions of gallons of contaminated water into the ground and that it will never seep into our aquifers or riverbeds is not realistic," Dr. Walter Tsou stated in September 2011 at a public hearing organized by the Citizens Marcellus Shale Commission.[487] Dr. Tsou is past president of the American Public Health Association and former Philadelphia health commissioner.

Dr. Tom Myers, a hydrologist, agrees. In a research study published in the April 2012 issue of *Groundwater*, Dr. Myers pointed out that fracking "could allow the transport of contaminants from the fractured shale to aquifers."[488] His research suggests that normal movement takes "tens of thousands of years to move contaminants to the surface, but fracking the shale could reduce that transport time to tens or hundreds of years" to contaminate the water supply.

A three-year study in Colorado, based upon methane samples from about 300 locations, revealed there was a broad level of water pollution caused by natural gas drilling.[489] A *ProPublica* summary of the September 2008 report noted:

"The researchers did not conclude that gas and fluids were migrating directly from the deep pockets of gas the industry was extracting. In fact, they said it was more likely that the gas originated from a weakness somewhere along the well's structure. But the discovery of so much natural fracturing, combined with fractures made by the drilling process, raises questions about how all those cracks interact with the well bore and whether they could be exacerbating the groundwater contamination."[490]

Biochemist Dr. Ronald Bishop suggested fracking to extract methane gas "is highly likely to degrade air, surface water and groundwater quality, to harm humans, and to negatively impact aquatic and forest ecosystems."[491] He noted that "potential exposure effects for humans will include poisoning of susceptible tissues, endocrine disruption syndromes, and elevated risk for certain cancers."[492]

Scientists at Duke University concluded, "methane contamination of shallow drinking water systems [is] associated with shale-gas extraction." The data and conclusions, published in the May 2011 issue of the *Proceedings of the National Academy of Sciences (PNAS)*, revealed not only did most drinking wells near drilling sites have methane, but those closest to the drilling wells, about a half-mile, had an average of 17 times the methane of those of other wells.[493]

A subsequent study by the research team, published in the July 2012 issue of *PNAS*, revealed that mineral-rich fluids deep in the Marcellus Shale are migrating to the surface, which disputes numerous industry claims that layers of impervious rock protect the escape of injected fluids in fracking. Although the scientists couldn't establish direct links between fluids used in fracking and the presence of methane and hazardous materials in aquifers and drinking water, they did find:

"the coincidence of elevated salinity in shallow groundwater with a geochemical signature similar to produced water from the Marcellus Formation suggests that these areas could be at greater risk of contamination from shale gas development because of a preexisting network of cross-formational pathways that has enhanced hydraulic connectivity to deeper geological formations."[494]

Nathaniel R. Warner, principal researcher, concluded that increased salinity of water in the Marcellus region "suggest that homeowners living in these areas are at higher risk of contamination from metals such as barium and strontium."[495]

In December 2011, following a three-year scientific investigation, the EPA issued a 121-page preliminary report that cited groundwater pollution as a direct result of fracking operations near Pavillion, Wyo.[496] The EPA determined there was contamination in 11 drinking wells, and that 169 gas wells had been drilled within the area that provides drinking water. Specifically rejecting alternate possibilities for the pollution, EPA scientists noted that at least 10 compounds—including the carcinogens benzene and 2-butoxyethanol—found in the water supply were specific to nearby fracking operations. The EPA also noted that potassium and chloride levels were higher than expected background levels. However, the EPA "was hamstrung by a lack of disclosure about exactly what chemicals had been used to frack the wells near Pavillion," according to *ProPublica,*[497] which noted that Encana, owner of the nearby wells, refused to give federal officials a list of chemicals it used in fracking operations but did claim there was no correlation to its mining and the pollution of the water supply.

Dr. Tom Myers ran his own tests at Pavillion, and determined that the EPA's "conclusion is sound." According to his report, published in April 2012, "Three factors combine to make Pavillion-area aquifers especially vulnerable to vertical containment transport from the gas production zone or the gas wells—the geology, the well design, and the well construction."[498] He did note, as had the EPA, that the area was unique because "the vertical distance between the water wells and fracking wells is much less at Pavillion than in other areas."[499]

Tests run by the U.S. Geological Survey several months after the EPA tests confirmed EPA findings.[500] The USGS tests showed the presence of diesel compounds, and high levels of ethane, methane, and phenol in drinking water. Encana continued to deny any relationship between fracking and contaminated water.

The EPA, with a $1.9 million grant from Congress, is investigating 10 sites, three in Pennsylvania, to determine if fracking contaminates water resources. That study is scheduled to be completed in 2014.

The public doesn't need to wait until 2014. "Even in the best case scenario, an individual well would potentially release at least 200 [cubic meters] of contaminated fluids," according to Daniel Rozell and Dr. Sheldon Reaven of the State University of New York at Stony Brook. Their study, published in August 2012, concludes, "This potential substantial risk suggests that additional steps be taken to reduce the potential for contaminated fluid release from hydraulic fracturing of shale gas."[501]

Christopher Portier, director of the National Center for Environmental Health, calls for more research studies that "include all the ways people can be exposed [to health hazards], such as through air, water, soil, plants and animals."[502]

PHOTO: Vera Scroggins

Flaring is used to burn off excess methane, which enters the atmosphere as a pollutant.

CHAPTER 8
Air Pollution

As if water pollution isn't bad enough, fracking operations also impact the air and increase greenhouse gas levels.

The oil and natural gas industry, according to the EPA, "is the largest source of emissions of volatile organic compounds (VOCs), a group that contribute to the formation of ground-level ozone (smog)."[503] The EPA estimates VOC emissions— about 200 times greater in fracked wells than conventional drilled wells—were 2.2 million tons in 2008,[504] the last year for data compilation. High concentrations of ozone can lead to shortness of breath, chest pain when inhaling, wheezing and coughing, throat irritation, asthma attacks, increased susceptibility to respiratory and pulmonary inflammation, and heart attacks, according to the American Lung Association.[505]

Researchers from the University of Colorado School of Public Health concluded that fracking may contribute to "acute and chronic health problems" from air pollution.[506] The research team headed by Dr. Lisa McKenzie had found "potentially toxic petroleum hydrocarbons in the air near the wells," citing benzene, ethylbenzene, toluene, trimethylbenzenes, and xylene. The four-person team pointed out these toxins:

"can adversely affect the nervous system with effects ranging from dizziness, headaches, fatigue at the lower exposures to numbness in the limbs, incoordination, tremors, temporary limb paralysis, and unconsciousness at higher exposures. Airborne chemicals from fracking could irritate the respiratory system and mucous membranes with effects ranging from eye, nose, and throat irritation to difficulty in breathing and impaired lung function."[507]

A team of researchers from Cornell University determined that leaked methane gas into the air from fracking operations could have a greater negative impact upon the environment than either oil or coal. Geochemist Dr. Robert W. Howarth, engineer Dr. Tony Ingraffea, and ecology researcher Renee Santoro, concluded:

> "The footprint for shale gas is greater than that for conventional gas or oil when viewed on any time horizon, but particularly so over 20 years. Compared to coal, the footprint of shale gas is at least 20% greater and perhaps more than twice as great on the 20-year horizon and is comparable when compared over 100 years. . . . The GHG [greenhouse gas] footprint of shale gas is significantly larger than that from conventional gas, due to methane emissions with flow-back fluids and from drill out of wells during well completion. . . . The large GHG footprint of shale gas undercuts the logic of its use as a bridging fuel over coming decades, if the goal is to reduce global warming."[508]

A comprehensive study by Dr. Marvin Resnikoff, a physicist, suggests that up to 30,000 cancer deaths in New York State from radon-contaminated gas mined in the Marcellus Shale, could be attributed to fracking operations. Dr. Resnikoff concluded:

> "The long-term environmental risks and public health concerns of radon in Marcellus Shale natural gas formations are far too serious to be ignored. The potential impacts of radon must not be swept under the rug. Nor should these impacts be sacrificed to short-term economic policies or to unrealistic and/or inaccurate assessments of the benefits of natural gas development in New York State."[509]

Natural gas producers in the Denver area are losing about 4 percent of their gas production into the atmosphere, about twice the earlier estimate, according to the National Oceanic and Atmospheric Administration (NOAA).[510] NOAA scientists, led by Dr. Gabrielle Petron, recorded considerably higher atmospheric levels of propane and butane than in Los Angeles or Houston, both of which have air pollution problems. Dr. Petron's team found that significantly increased levels of methane in the air came from oil and gas production. She also found significant levels of the carcinogen benzene. Analyzing thousands of data readings, the team found benzene in the atmosphere was between 385 and 2,055 metric tons per year, significantly higher than the previously reported estimates of 60 to 145 tons released each year.[511]

Infrared videography revealed air pollution near 11 well sites in Pennsylvania, Maryland, and West Virginia. The Chesapeake Bay Foundation, which hired Dr. Howarth to document the presence of airborne pollution, said the video "establishes that the industry is not sufficiently limiting the amount of leaks from drilling and processing operations."[512] Dr. Howarth filmed 15 plants, four of which had no evidence of airborne pollution. While conceding that natural gas is cleaner than coal and oil, the Foundation also told the Associated Press in November 2011, "The alarming rate at which extraction activities have increased in the bay watershed gives us great pause as we attempt to understand the full implications."[513]

A *ProPublica* investigation by Abrahm Lustgarten revealed cases where individuals living near wells developed health problems far greater than expected if they had lived away from the wells. In one case, a Colorado woman living less than a half-mile from a well was forced to wear an oxygen mask every time she left her house. According to *ProPublica*, which talked with residents in Pennsylvania, Colorado, and Texas, "most common complaints are respiratory infections, headaches, neurological impairment, nausea and skin rashes. More rarely, they have reported more serious effects, from miscarriages and tumors to benzene poisoning and cancer."[514]

Janet McIntyre, who lives about 1,500 feet from two well pads in Connoquenessing Twp., Butler County, Pa., told Matt Walker of the Clean Air Council:

> "There's this blue haze that emits over this whole area here. It's just this blue haze— it's everywhere. You can't be outside more than 10 minutes. You get massive headaches, your eyes are red, your lips become tinny-tasting. I lived here 20 years and never had an issue until they came to town. If it's hot and you're outside, it feels like your skin is burning and is going to crawl off—it's bad."[515]

Vapor clouds released during fracking operations by Chesapeake Energy possibly led to illness among residents of Arlington, Texas, according to an investigation by Peter Gorman for the *Fort Worth Weekly*.[516] One resident, Jane Lynn, told Gorman that in December 2011, "I was literally in the vapor cloud, and the odor was overwhelmingly strong. Within minutes there was tightness in my chest. Later that day I began getting heart palpitations—which I've only gotten previously when exposed to gas escaping from a well site."[517] Ranjana Bhandari said she was in traffic when fumes from a well began to affect her sinuses. "The symptoms lasted for hours and then returned when I was stuck at a railway crossing in the vicinity a few days later," Bhandari said.[518] Jean Stephens said after walking out of her store she was overcome by nausea and breathing difficulties from a vapor cloud. "It was just a severely nauseating foul odor [that] I'd never smelled . . . before," she said.[519] The response from Chesapeake, according to the *Weekly*, was the vapor clouds were only clean steam.

Mayor Calvin Tillman of the small Texas town of DISH, surrounded by wells and compressor stations, told the *Fort Worth Weekly*, "Some days you can hardly breathe anywhere in DISH. . . . We knew something was terribly wrong." However, when the residents complained to state agencies, the agencies "basically asked the companies to investigate themselves, and they came up clean every time."[520]

And so the residents voted to commit $10,000 of their annual $70,000 budget to fund a study of air analysis. According to the report, chemist-microbiologist Dr. Wilma Subra studied "fugitive emission sources of hazardous air pollutants emanating from the oil and gas sector include emissions from pumps, compressors, engine exhaust and oil/condensate tanks, pressure relief devices, sampling connections systems, well drilling (hydraulic fracturing), engines, well completions, gas processing and transmissions as well a mobile vehicle transportation emissions."[521] The testing, says Dr. Subra:

> "confirmed the presence in high concentrations of carcinogenic and neurotoxin compounds in ambient air near and/or on residential properties. The compounds in the air indicate quantities in excess of what would normally be anticipated in ambient air in an urban residential or rural residential area.

Many of these compounds verified by laboratory analysis were metabolites of known human carcinogens and exceeded both Short-term and Long-term effective screening levels . . ."[522]

Among the known carcinogens released into the air, presumably by the wells and compressor stations were benzene, carbon disulfide, naphthalene, trimethylbenzene, and xylene toxins. Dr. Alisa Rich, an environmental scientist and president of the research firm that conducted the study, told the *Fort Worth Weekly*, "it's a toxic soup out there."[523]

The response by the Texas Pipeline Association was to deny the scientific conclusions and question the study's methodology. In March 2012, Calvin Tillman resigned as mayor and moved out of DISH, citing health concerns for his family. Tillman said his two sons, ages 4 and 7, were having respiratory problems and nosebleeds, which he attributed to the air pollution from the gas industry.[524]

Pollution from Wells and Compressor Explosions

In June 2010, a "blowout" at a well owned by EOG Resources spewed about 35,000 gallons of toxic chemicals and about one million gallons of hydrofracking fluid into the air for 16 hours in the Moshannon State Forest in central Pennsylvania. The state DEP fined EOG about $400,000 and shut down operations for 40 days.[525]

Three persons in Pearsall, Texas, were injured after an explosion and fire at a wastewater disposal well in January 2012. The Railroad Commission said the cause of the explosion was welders who were working near a truck that was unloading the water into a collection tank. "Sparks from the welding may have ignited vapors around the storage tank, causing the explosion," according to an official statement.[526] While human error may have caused the explosion, the reality is that the wastewater had enough volatile fumes that could have impacted health and safety even if welding sparks hadn't caused the problem.

Three Chesapeake Energy workers were burned from an explosion and subsequent fire at a compressor station near Avella, Pa., in February 2012. The Pennsylvania DEP attributed the cause to ignition of escaping vapors from five steel tanks that held wet gas, also known as condensate, a product in natural gas. "If you've ever been to one of those well sites, the fumes are very bad," Avella fire chief Eric Temple told the *Pittsburgh Tribune-Review*.[527] (Between Jan. 1, 2009, and August 1, 2012, Pennsylvania approved more than 500 permits for compressor stations, according to Clean Air Council data.

A natural gas compressor in Springville Twp., about 30 miles northwest of Scranton, Pa., exploded March 29, 2012, destroying the sheet metal roof of the compressor station and sending flames and thick clouds of black smoke into the air. The compressor station is owned and operated by Williams Partners of Tulsa, Okla. The DEP initially reported the cause was possibly a leak in a gas line; it later reported the cause was a valve that was left open during maintenance work.[528] There were no injuries. The compressor station can process about 365 million cubic feet of gas per day, and then send it along two 24-inch diameter pipelines.[529] One day after the explosion, and against orders by the DEP, Williams again was sending natural gas into the pipelines.[530] An investigation of federal records by Natural Gas Watch revealed the company had several violations in its multi-state operation and may have been responsible for a natural gas explosion that destroyed two homes and injured five persons near Appomattox, Va., Sept. 25, 2008.[531]

PIPELINE REGULATION

An inspector from the Public Utilities Commission was on scene within two hours of the explosion in Springville Twp., but did not have jurisdiction. A four-part series by the *Philadelphia Inquirer* in December 2011 had revealed that no state or federal agency has jurisdiction over pipelines in Class 1 rural areas, nor are operators required to report any incidents, including property damage, injuries, or deaths associated with those pipelines.[532] Pennsylvania's Gas and Hazardous Liquids Pipeline Act,[533] which became law in December 2011, includes oversight of classes 2–4, but specifically excludes Class 1 pipelines. A Class 1 location is any area with "10 or fewer buildings intended for human occupancy within 220 yards of the centerline of the pipeline," according to the Pipeline and Hazardous

Materials Safety Administration (PHMSA). About 1,300 miles of Pennsylvania's natural gas pipelines are Class 1 pipelines.

"For decades, the gas industry has fought hard to protect that exemption, defeating repeated attempts by Congress and safety advocates to change it," the *Inquirer* reported shortly after the explosion in Springville Twp.[534] One of the major problems in excluding Class 1 pipelines is that "[t]he vast majorities of Class 1 pipelines are and will be located in rural areas where first responders will most likely be untrained on how to handle an ignited well," according to *Jurist*, a web-based newsletter about legal issues.[535]

Regulating Class I pipelines is "at the bottom of the state's priority list," said Patrick Henderson, energy executive for the Corbett Administration.[536]

About half of the nation's 2.6 million miles of pipelines are at least 50 years old; corrosion, according to *ProPublica*, is responsible for between 15 and 20 percent of deaths, injuries, or property damage.[537] More than 150 incidents a year involve large natural gas transmission lines and the smaller distribution lines.[538]

Eight persons died, 50 were injured, and 38 houses destroyed from an explosion in a 30-inch diameter natural gas pipe in San Bruno, Calif., Sept. 9, 2010.[539] The explosion, which led to flames more than 1,000 feet high,[540] also left a 40-foot deep crater that measured 167 feet long by 26 feet wide at the surface.[541] Because there were no automatic shut off valves on the pipe, which was laid in 1956, it took workers almost 90 minutes to manually shut off the fire.[542] An independent investigation ordered by the California Public Utilities Commission later revealed that Pacific Gas & Electric (PG&E) had diverted more than $100 million for safety improvements to executive bonuses and increased stockholder dividends. The report noted that PG&E gave a "low priority" to safety, and that its "focus on financial performance [was] well outside industry practice— even during times of corporate austerity programs."[543]

More than 900 persons were evacuated, and thousands were ordered to remain in their houses with doors and windows closed, when three explosions followed by a black plume rose to about 1,000 feet from the Chevron Oil Refinery plant in Richmond, Calif., in the late evening of Aug. 6, 2012.[544] That cloud of toxic hydrocarbons began drifting over cities in the San Francisco Bay area.[545] A leaking eight-inch diameter pipe, first placed in the plant in the 1970s, was believed to be the source of the plume and fire. In addition to five employees who were injured, more than 5,700 residents sought medical evaluation and treatment; about 160,000 persons live in the affected area.[546]

"We have been fighting Chevron since I was elected," Mayor Gayle McLaughlin told NPR's *Democracy Now!* With a new Council membership, McLaughlin says "Chevron has no longer dominated City Hall,"[547] part of the reason why Richmond had previously accepted most of what Chevron was dealing.

Antonia Juhasz, author of several books about the oil industry, reported in the *Los Angeles Times* that Chevron, the state's largest employer, has a record that leads to incidents that threaten the public health and welfare. "Rather than use its $27 billion in 2011 profits to run the cleanest, safest and most transparent refinery possible," said Juhasz, "Chevron operates a refinery that is in constant violation of federal and state law and a daily threat to the health and safety of its workers and neighbors." According to Juhasz:

"Since at least April 2009, the refinery has been in non-compliance of the Clean Water Act and the National Pollutant Discharge Elimination System in every quarter but one. Until July 2010, the refinery had been in 'high-priority violation' of Clean Air Act compliance standards, the most serious level of violation noted by the EPA, since at least 2006. Under constant pressure from community organizations, Chevron has been assessed hundreds of thousands of dollars in penalties for repeated Clean Air Act violations—nearly 100 citations in just the last five years, including 23 in 2011 alone. A 2008 study by UC Berkeley and Brown University researchers concluded that the air inside some Richmond homes was more toxic than that outside because of harmful pollutants from the refinery being trapped indoors.

"The Contra Costa County Health Services Department lists the residents of Richmond as one of the 'most at-risk groups' in the county: They are hospitalized for chronic diseases at significantly higher rates than the county average, including for female reproductive cancers, which are more than double the county rate. Chevron is one of four refineries in Contra Costa County where nearby incidence of breast, ovarian and

152

prostate cancers are the second highest in California, and where nearby residents suffer higher rates of asthma, childhood asthma and asthma-related deaths."[548]

An explosion near Sissonville, W. Va., in December 2012 that destroyed four houses and about 800 feet of Interstate 77[549] was probably the result of sudden pressure drop in a six-foot section of pipe that had worn to 0.078 inches of thickness, according to the National Transportation Safety Board (NTSB).[550]

The NTSB also determined that it took more than an hour for Columbia Gas Transmission employees to manually shut off the gas. For almost two decades, the federal agency had urged, but could not require, companies to install automatic shut-off valves. Jim Hall, NTSB chair from 1994 through 2000, told the Associated Press following the Sissonville explosion, "The companies' attitude is, in many cases, unless it's required, they're not going to do it."[551] A law signed by President Obama in 2012 to require automatic shut-off valves has not been implemented. The AP reports that the Congressional mandate, still being fine-tuned, would require such valves only when it's "economically, technically and operationally feasible." Thus, existing pipes probably will not be retrofitted for automatic shut-off valves, but new pipes might be.

A one year investigation by the Government Accountability Office (GAO), initiated by the Senate Committee on Commerce, Science, and Transportation, revealed that the PHMSA regulates only about 20,000 of 200,000 miles of natural gas gathering pipelines (such as those at condensors) and only about 4,000 of the estimated 30,000–40,000 miles of hazardous liquid gathering pipelines. The report, published in March 2012, noted:

"In response to GAO's survey, state pipeline safety agencies cited construction quality, maintenance practices, unknown or uncertain locations, and limited or no information on current pipeline integrity as safety risks for federally unregulated gathering pipelines.

"Our survey revealed that only 3 of the 39 state agencies reported that they collect and analyze comprehensive pipeline spill and release data on federally unregulated pipelines."[552]

The GAO recommended the Department of Transportation should "Collect data from operators of federally unregulated onshore hazardous liquid and gas gathering pipelines [and] establish an online clearinghouse or other resource for states to share information on practices that can help ensure the safety of federally unregulated onshore hazardous liquid and gas gathering pipelines."[553]

Pollution from Traffic Accidents

In addition to the normal diesel emissions of trucks and trains, there are numerous incidents of leaks, some of several thousand gallons, much of which spills onto roadways and into creeks, from highway accidents of tractor-trailer trucks carrying wastewater and other chemicals.[554]

Each day, interstate carriers transport about five million gallons of hazardous materials.[555] Not included among the daily 800,000 shipments are the shipments by intrastate carriers, which don't have to report their cargo deliveries to the Department of Transportation. "Millions of gallons of wastewater produced a day, buzzing down the road, and still nobody's really keeping track," Myron Arnowitt, the Pennsylvania state director for Clean Water Action, told Aaron Skirboll of AlterNet[556]

One of the most serious incidents occurred July 11, 2012, when three tank cars on a Norfolk Southern train carrying 90,000 gallons of ethanol exploded near Columbus, Ohio. The explosion led to the evacuation of about 100 residents within a mile of the accident.[557]

A 4,000 gallon spill of hydrochloric acid near Canton, Pa., July 4, 2012, resulted in pollution to a tributary of Towanda Creek.[558]

Several thousand gallons of wastewater spilled into a storm drain that empties into Pine Creek, near Jersey Shore, Pa., Sept. 26, 2012, when a truck hauling 4,600 gallons of wastewater crashed on the way to a gas well.[559]

On the evening of Aug. 9, 2012, a Halliburton truck carrying 4,000 gallons of hydrochloric acid pulled off Interstate 80, and into a convenience store market/gas station near New Columbia, Pa., after the driver noticed a leak. "There was a huge plume in the air and it was just getting bigger and bigger," resident Amanda Friend told the Sunbury (Pa.) *Daily Item.*[560] Customers immediately fled when they saw the plume, which eventually engulfed the store, according to the *Daily Item.* HazMat teams detected a hole that had leaked at least 250 gallons of the acid. Larry Maynard, White Deer Twp. EMA director, told the *Daily Item* that because of the increased truck traffic due to fracking operations, EMA officials trained "many times" to deal with the probability of hazardous materials leaks. "If you want a shock, just park along I-80 and watch what goes by," Maynard said.[561] The problem of a toxic fume that would affect several body systems was diminished only because a westerly wind blew the plume away from the village.

The gas leak could have been detected and the problem resolved much quicker had the truck been equipped with a leak-sensor camera, which detects hazardous materials leaks within 15 microseconds. In a letter-to-the-editor to the *Daily Item,* Flora Eyster, who lives near New Columbia, suggested:

"Not only should the DEP and others require and monitor compliance, but every truck hauling these materials should be required to 1. Install and use this device 2. Be trained in a legally-required response and 3. Be REQUIRED to follow certain communication paths and immediate call-in response from drivers. The first call should always be to a central qualified source who fields responses and information to all involved—especially on the ground and those immediately in the response area"

"In Pennsylvania, workers have reported contact with chemicals without appropriate protective equipment, inhalation of sand without masks, and repeated emergency visits for heat stroke, heat exhaustion, yet many of the medical encounters go unreported," says Dr. Pouné Saberi, a public health physician affiliated with *Protecting Our Waters.*[562]

Other problems are truck reliability and driver fatigue. About 40 percent of all gas and oil trucks inspected by the Pennsylvania State Police between January 2009 and February 2012 had enough mechanical problems that they were ordered off the road, according to reporting by Ian Urbina in the *New York Times.*[563] Urbina also noted that because of work shifts that often exceed the 14 hours limit to commercial truckers, gas industry truckers are putting in work shifts of as many as 20 hours, leading not only to fatigue but accidents as well; more than 300 oil and gas workers were killed in highway crashes between 2002 and February 2012.

A truck driver who was afraid to identify himself, told a community meeting in Heber Springs, Ark., in April 2011, more truths:

> "Now, these drivers will not get up and stand up . . . because their wages are better, they get $15 or $20 an hour; they think they are in high cotton. But just the same, anything I had to say, when they told me go clean a frac truck out, I said under OSHA regulations I've got to have a hazmat suit on. They laughed me out of it.
>
> "It's a culture of fear that's in the oil industry right now, similar to the coal miners in West Virginia [that] told ya', 'Well we really did not want to tell you the truth about how bad the coal mine is.' It's bad out there, too. A majority of people at a rally last weekend [said] . . . 'We're for the Industry! Yeah, yeah, yeah! Because we're making money.' But I guarantee you what's going on, they're dumping illegally . . .
>
> "This industry is a bunch of liars. Don't believe what they have to say. And until we find out if this stuff is safe, we need to have a moratorium."[564]

It may be months or years before the drivers from Halliburton or the unnamed water tanker driver in Arkansas learn the extent of possible injury or diseases caused by industry neglect. For José Lara of Rifle, Colo, the effects no longer matter. Lara, an employee of Rain for Rent died at the age of 42 in August 2011 from pancreatic and liver cancer. Lara's job was to power-wash wastewater tanks owned by natural gas companies. In a six-hour deposition three months before he died, Lara said he was never provided a respirator or protective clothing. "The chemicals, the smell was so bad. Once I got out, I couldn't stop throwing up. I couldn't even talk," Lara said in his deposition, translated from the Spanish.[565]

John Dzenitis of KREX-TV reported:

> "Both the industry and the Colorado Oil and Gas Conservation Commission, the state's agency meant to protect public

156

health and regulate oil and gas, have denied the existence of high levels of hydrogen sulfide in Colorado. In 1997, the Colorado Department of Public Health and Environment wanted to monitor for hydrogen sulfide at oil and gas facilities after they were designated as confirmed sources of the deadly gas by the EPA.

"The COGCC stepped in and told them not to, claiming there were no elevated levels in the state. The public health department listened, and tells us they haven't pursued any monitoring of hydrogen sulfide at oil and gas facilities since."[566]

No matter what Colorado officials did or didn't do, it didn't keep the Occupational Safety and Health Administration (OSHA) from issuing Rain for Rent nine violations for exposing Lara to hydrogen sulfide and not adequately protecting him from the effects of the cyanide-like gas. Nevertheless, hydrogen sulfide monitoring is exempt from regulations of the Clean Air Act.[567] That may change if a bill survives Congressional inaction. Reps. Jared Polis (D-Colo.), Maurice Hinchey (D-N.Y.), and Rush Holt (D-N.J.) introduced a resolution (HR 1204[568]) that would close that loophole. The bill was introduced in March 2011 but under Republican leadership was buried in the Committee on Energy and Power.

Health Problems with Sand

As much as three million pounds of "frack sand" is needed for each well.[569] The best sand, a fine white sand with crystalline silica, is mined in the Midwest, primarily in Minnesota and Wisconsin. Increased train traffic brings the sand to the Barnett, Marcellus, and other shales; trucks bring the sand from rail terminals to the well.

The increased sand mining, similar to coal field strip mining, has destroyed agricultural and scenic land, and has led to the land "being ground up and shipped away, only to become toxic, radioactive waste somewhere else," journalist Pilar Gerasimo reported in the *Dunn* (Wisc.) *County News*.[570]

Because of health and environmental concerns raised by Gerasimo, several communities established temporary moratoriums on sand mining; Dunn County had placed a six month moratorium, beginning January 2012, on non-metallic mining, and then extended it in July for three more months.[571] The moratorium ended at the end of October, but the county says it has not received any requests to drill nonconventional wells or for sand mining.

Sand is also the key to methane migration into aquifers and well water because its function is to keep the fractures open to allow methane to flow from the shale.

The effects of chemicals and the sand used in the fracking process so concerned the AFL-CIO that in May 2012 three senior health and safety officials wrote to divisions within the Department of Labor to express the concern of the 11 million member federation.[572] Referring to a study by the National Institute for Occupational Health and Safety (NIOSH), which had identified high levels of worker exposure to crystalline silica, the AFL-CIO observed:

"Many of these exposures were well in excess of permissible and recommended levels, putting workers at risk of silicosis, lung cancer and other diseases. These findings coupled with concerns about health risks posed by chemical additives used in the fracking process and the well-documented safety hazards in this industry warrant immediate attention and action."[573]

The AFL-CIO urged the federal government to consider that "the development of new energy sources, and exploration of existing energy sources, must be done safely without putting workers in danger," and suggested "effective regulation and oversight."

Following its own study and the AFL-CIO concerns, NIOSH issued a Hazard Alert to the effects of crystalline silica, noting there were seven primary sources of exposure during the fracking process, all of which could contribute to workers getting silicosis, the result of silica entering lung tissue and causing inflammation and scarring.[574] Excessive silica can also lead to kidney and autoimmune diseases, lung cancer, tuberculosis, and Chronic Obstructive Pulmonary Disease (COPD). In the Alert, NIOSH pointed out that its studies revealed about 79 percent of all samples it took in five states exceeded acceptable health levels, with 31 percent of all samples exceeding acceptable health levels by 10 times. NIOSH issued several suggestions of how the industry could monitor air samples and worker health, control dust, and provide respiratory protection. However, the Hazard Alert is only advisory; it carries no legal or regulatory obligations.

New EPA Regulations

In January 2009, WildEarth Guardians and the San Juan Citizens Alliance had sued the EPA to force compliance with the Clean Air Act of 1970. That Act requires the EPA every eight years to review standards for industrial categories that could cause or contribute to air pollution and endanger public health and the environment. The last time the EPA conducted a study was in 1985. In issuing a consent decree in February 2010, the U.S. District Court for the District of Columbia ordered the EPA to present a final set of regulations no later than Jan. 31, 2011.[575] The Court later granted EPA requests for extensions.

On April 17, 2012, with strong input from the natural gas industry and having consulted environmental groups, the EPA filed the regulations on the last date allowed by the Court.[576] The regulations require natural gas drillers to install special equipment to separate gas and hydrocarbons in the flowback. The EPA says the rules "are expected to yield a nearly 95 percent reduction in VOC [volatile organic compounds] from more than 11,000 new hydraulically fractured gas wells each year" and by 25 percent the methane gas released into the air.[577] The EPA believes the new regulations will reduce VOCs, which can cause ground level smog, by 190,000–290,000 tons, air toxins by 12,000–20,000 tons, and methane by 1.0–1.7 tons.[578]

However, in a major concession to the natural gas industry, the rules will not go into effect until January 2015, six years from the time the suit was first filed. "The concession only promotes wasteful drilling," said Jeremy Nichols, climate and energy program director of WildEarth Guardians.[579]

Water trucks on Route 6 near Wysox, Pa.

PHOTO: Diane Siegmund

CHAPTER 9:
Impact of Diesel Fuel and Noise Pollution

Equally important to understanding the health hazards from air, water, and ground pollution are questions about the impact of diesel fuel in fracking solutions and from transportation. Diesel fuel is soluble in water and can pollute ground, water, and air. It contains a number of toxic chemicals and carcinogens, including benzene, toluene, ethylbenzene, and xylene, commonly known as BTEX, which has significant negative effects upon health.[580] The industry illegally uses diesel fuel in the fracking process.

In 2003, three major natural gas corporations had signed an agreement with the EPA not to use diesel fuel in the fracking process. A Congressional investigation later determined that BJ Services and Halliburton violated that agreement between 2005 and 2007; Schlumberger did not.[581]

Both the Safe Water Drinking Act and the Energy Policy Act of 2005 prohibit the use of diesel fuel in the fracking mixture. However, an investigation by EcoWatch revealed that between January 2011 and August 2012 the natural gas industry used kerosene and diesel fuels 1 and 2 as part of the fracking fluid mix on 448 separate occasions.[582] EcoWatch determined that wells in four states had more than 90 percent of all violations— Arkansas (171), Texas (142), New Mexico (57), and Pennsylvania (45).

"The list of toxic chemicals exempted by the Halliburton loophole is staggering, but to find that the one item still restricted is nonetheless being used without regulation or consequence is unacceptable," EcoWatch observed, and argued, "Not only do these health hazards raise concern about injection through groundwater supplies to shale layers deep beneath the earth, but also [by] air transmission through flaring and fugitive emissions."[583] The only good news from the EcoWatch investigation is "the frequency of diesel fuels' use in fracking . . . appears to be on the overall decline."[584]

A federal law enacted in 2010 requires the use of ultra-low sulfur diesel fuel (ULSF) in transportation,[585] which has resulted in a "cleaner" fuel, and more efficient engines. This has resulted in a significant reduction in particulate emissions and nitrogen oxide. However, the Diesel Technology Forum acknowledges only about "one-third of all the heavy-duty commercial trucks on the road in the U.S. are 2007 or later model year and have the most sophisticated emissions control technology."[586]

Although technology has eliminated much of the pollution from diesel-fuel transportation, another problem exists. "Diesel engines have long been associated with refinery and drilling rig disasters because they can overheat and rev to the point of explosion when their intake valves suck in hydro-carbon vapor," according to the *San Francisco Chronicle*.[587] Diesel fuel, if spilled onto highways, is also more difficult to wash away than gasoline and will often leave a greasy slick that becomes a hazard to driving conditions.

Dr. Ronald Bishop, professor of biochemistry at the State University of New York at Oneonta, believes "intensive use of diesel-fuel equipment will degrade air quality [that could affect] humans, livestock, and crops."[588]

Barb Harris, a specialist in environmental toxins, points out, "Pre-natal exposure to diesel fuel combustion products is a known risk for low birth weight [which is] a major indicator of child health, and is associated with multiple health and learning issues throughout life."[589]

The gas boom has led to significantly increased truck and train traffic; the trucks, as many as 200 a day per well, each one with two 100–200 gallon fuel tanks, bring water to the site and then remove the wastewater; dozens of trucks bring supplies; trains bring sand and other chemicals.

The New York State Department of Environmental Conservation estimates about 4,000 truck trips per well during the first 50 days of development is needed for each gas well that uses the fracking procedures;[590] this is almost three times more than for conventional wells drilled without the use of fracking.

For short line railroads, activity in the Marcellus Shale, beginning about 2008, has been a boost in productivity and profits and, according to *Marcellus Drilling News*, "the revival of short line railroads.[591] The expansion of gas drilling "has resulted in a spike in demand primarily for 'frac' sand, but also pipe, chemicals and other materials," reported *Progressive Railroading*.[592] The monthly trade magazine also forecast that a possible increase in fracking operations "of about 43 percent over the next two years" [2012 and 2013] will have significant impact on rail traffic. The North Shore Railroad System, which owns six short lines in northeastern Pennsylvania, reported a 40 percent increase in business in the first quarter of 2012 over a year earlier.[593] CSX, the largest freight railroad on the east coast, in 2011 purchased 900 more cube hopper cars designed specifically to move frack sand.[594] The railroad reported it carried more than 12,000 carloads of sand into the Marcellus Shale region in 2011, an increase of 40 percent from the previous year.[595] Norfolk Southern, with track in the Marcellus Shale, in 2012 spent about 14 percent of its entire capital expenditures for cube hoppers.[596]

A Pennsylvania law[597] implemented in 2009 forbids trucks to idle more than five minutes per hour. "Idling of these heavy-duty engines produces large quantities of dangerous air pollutants that can be particularly harmful to young children, the elderly and people with respiratory problems, such as asthma, emphysema and bronchitis," John Hanger, the state's DEP secretary, told *The Trucker* magazine in 2009.[598] Pennsylvania has not enforced that law at natural gas sites since at least 2010 when Tom Corbett became governor and appointed Michael Krancer to replace Hanger as DEP secretary.

"We need to know how diesel fuel got into some people's water supply," says Diane Siegmund, a clinical psychologist from Towanda, Pa. "It wasn't there before the companies drilled wells; it's here now," she says. Siegmund is also concerned about contaminated dust and mud. "There is no oversight on these," she says, "but those trucks are muddy when they leave the well sites, and dust may have impact miles from the well sites."

For decades, the industry wasn't concerned about diesel fumes. "They're temporary sources, and they are not causing accumulated effects," said Nicholas (Corky) DeMarco, executive director of the West Virginia Oil and Natural Gas Association.[599] The problems are magnified by trucks idling at sites, and by leaks and spills while on the road.

However, the industry is beginning to take advantage of a natural gas glut from overdrilling, and the resulting lower costs, to retrofit pumps and rigs to run on natural gas. The largest manufacturing companies—Baker Hughes, Halliburton, and Schlumberger—began retrofitting their equipment in 2012.[600] Their decisions had little to do with the desire to protect public health, and everything to do with the desire to reduce costs and maximize profits.

Noise Pollution

The constant noise level and artificial lighting have adverse effects upon both humans and wildlife.

"Noise from excavation, earth moving, plant and vehicle transport during site preparation has a potential impact on both residents and local wildlife, particularly in sensitive areas," according to 292-page report for the European Commission. Based upon extensive field analysis, the report states:

> "Well drilling and the hydraulic fracturing process itself are the most significant sources of noise. Flaring of gas can also be noisy. For an individual well the time span of the drilling phase will be quite short (around four weeks in duration) but will be continuous 24 hours a day. The effect of noise on local residents and wildlife will be significantly higher where multiple wells are drilled in a single pad, which typically lasts over a five-month period. Noise during hydraulic fracturing also has the potential to temporarily disrupt and disturb local residents and wildlife. Effective noise abatement measures will reduce the impact in most cases, although the risk is considered moderate in locations where proximity to residential areas or wildlife habitats is a consideration.
>
> "It is estimated that each well-pad (assuming 10 wells per pad) would require 800 to 2,500 days of noisy activity during pre-production, covering ground works and road construction as well as the hydraulic fracturing process. These noise levels

would need to be carefully controlled to avoid risks to health for members of the public."[601]

Noise pollution "is not only an environmental nuisance but also a threat to public health," says Zsuzsanna Jakob, regional director for the World Health Organization. According to Drs. NAA Castelo Branco, an occupation medical researcher; and Mariana Alves-Pereira, a physicist and environmental scientist:

> "Vibroacoustic disease (VAD) is a whole-body, systemic pathology, characterized by the abnormal proliferation of extra-cellular matrices, and caused by excessive exposure to low frequency noise (LFN) . . .
> "In both human and animal models, LFN exposure causes thickening of cardiovascular structures. Indeed, pericardial thickening with no inflammatory process, and in the absence of diastolic dysfunction, is the hallmark of VAD. Depressions, increased irritability and aggressiveness, a tendency for isolation, and decreased cognitive skills are all part of the clinical picture of VAD."[602]

Pennsylvania law gives natural gas companies authority to operate compressor stations continuously at up to 60 decibels,[603] the equivalent of continuous conversation in restaurants. However, most other states don't have limits. Compressor stations don't have to be enclosed, which would reduce noise levels.

Calvin Tillman, the former mayor of DISH, Texas, says the natural gas industry "seem[s] to think it's okay to put a compressor in that constantly emits 100 decibels of noise," louder than a lawnmower, and the industry thinks "that that's okay, and that's well within their rights to do so, and they don't give us any relief." The industry, says Tillman, "can make a compressor station that sounds very similar to the air conditioner on your home. . . . But they won't unless you make them. And we've had to fight and fight and fight, and we've had to spend money."[604]

"There is noise twenty-four [hours], seven [days a week]," Jennifer Palazzolo, a resident of Erie, Colo., told the *Colorado Independent*.[605] She said the noise from the trucks going to and from the Canyon Creek well site "is waking people up at night and this is only the beginning." The Denver–Julesburg Basin, northeast of Denver, is the site of hundreds of well pads, many within sight of schools. One well pad, reported the *Independent*, is within a hundred yards of the Aspen Ridge Charter School; Erie High School "is also surrounded by drill pads."[606]

Matt Pitzarella of Range Resources acknowledged "There are lights, some noises, some road dust." However, he claimed, "[W]ithin a year it's all gone and everything is put back together."[607]

Calvin Tillman, Jennifer Palazzolo, and thousands of others may not agree.

PHOTO: Robert Donnan

A convoy of trucks in Dryden, Pa., block traffic. It's a common scene throughout the Marcellus Shale region.

CHAPTER 10:
Effects Upon Agriculture, Livestock, and Wildlife

Building roads, pipelines, compressors, and drilling pads lead not only to decreased agriculture production and the destruction of the environment, but also to numerous accidents that impact the health of people and animals.

Destroying the Food Supply

About 360 million years ago, the Bakken shale began forming in the area now known as South Dakota, Montana, and Saskatchewan.[608] It is now about 200,000 square miles, lying between 4,500 and 7,500 feet below the surface of the earth.[609]

Oil in the shale was discovered in 1953;[610] however, because the shale is only 13 to 140 feet thick, using conventional drilling methods were marginally profitable. And then came the process of hydraulic horizontal fracturing, which allowed the energy companies to drill deep into the earth, and then snake their pipes and tubes horizontally for as much as a mile.

Energy company landmen, buying land and negotiating mineral rights leases, became as pesky as aphids in the wheat fields. However, the landmen didn't have to do much sweet talking with the farmers, many of whom were hugging bankruptcy during the Great Recession. The farmers yielded parts of their land to the energy companies in exchange for immediate income and the promise of future royalties. Even if the farmers didn't want to lease part of their land, many didn't have a choice—others, not them, owned the sub-surface mineral rights. The first leases went for as little as $10 an acre plus royalties for five years. But, all leases had a termination clause; if the company didn't begin active drilling within those five years, the farmer or landowner would again hold mineral rights. To avoid having to renegotiate the lease at a substantially higher price, as landowners became more sophisticated in the value of their land, the energy companies drilled shallow wells, taking just enough oil and gas to show activity. By November 2012 there were 7,791 wells in North Dakota.[611]

In 2006, oil production in the North Dakota fields was about 2.2 million barrels;[612] the next year, it was about 7.4 million barrels. Recoverable oil and gas continued to increase. Energy companies mined about 290 million barrels of oil in 2012,[613] and production is expected to increase to more than 360 million barrels in 2013.[614] There may be 18 billion barrels of recoverable oil, just in North Dakota alone.[615]

Drilling for oil also yields natural gas; there are about two trillion barrels of natural gas in the shale.[616] North Dakota has 16 natural gas processing plants, with a capability to produce about 800 million cubic feet of gas a day, according to Justin J. Kringstad, director of the state's Pipeline Authority.[617] "That represents a tripling of gas processing capacity in just the last five years. By the end of 2012, gas processing and transport capability [rose] to just over 1.1 billion cubic feet per day," Kringstad told David Fessler, an energy investment analyst.[618]

The oil and gas boom in North Dakota has led to the lowest unemployment rate in the nation.[619] While the nation's rate hovered about 7.9 percent,[620] North Dakota averaged 3.1 percent unemployment in October 2012.[621]

Energy company workers are earning annual incomes in the high five figure range; even entry-level jobs in local businesses are paying a minimum wage of $12 an hour in some parts of the state.[622] Businesses in the mining region that may have been marginally profitable at one time are showing double-digit profits. As business takes advantage of the housing boom, the cost of living has increased significantly.[623] The boom has also led to a housing shortage; hundreds of workers are living in tents, RVs, campers, or their cars, often paying as much as $1,200 a month for lot rentals.[624] The rent for one bedroom apartments average about $1,500 a month; two bedroom apartments, if available, are renting for twice that; motels and hotels, with inflated rates, are full. The boom has also led to significant depletion of the water supply, and has strained the capacity of local government to provide adequate services.[625]

Even if the impossible occurred, and there were no workplace injuries and deaths, even if there was no damage to the environment and to public health, there is one problem that can't be solved. The Bakken Shale lies directly below one of the most fertile wheat fields in the United States. To get to the shale, companies must destroy the fields.

Those fields are primarily amber durum wheat. High in protein and one of the strongest of all wheats, amber durum is a base for most of the world's food production. It is used for all pastas, pizza crusts, couscous, and numerous kinds of breads. Red durum, a variety, is used to feed cattle. North Dakota farmers in late Summer harvest about 50 million bushels (about 1.4 million tons) of amber durum,[626] almost three-fourths of all amber durum harvested in the United States.[627] About one-third of the production is exported, primarily to Europe, Africa, and the Middle East.[628] The destruction of the wheat fields, from a combination of global warming and fracking, will cause production to decline, prices to rise, and famine to increase.

The destruction of agricultural lands is not unique to North Dakota. Wherever energy companies drill there is a loss of agriculture.

Canada's National Farmers Union (NFU) has called for a moratorium on fracking. Jan Slomp, coordinator of the NFU in Alberta, told the media, "We are in the heart of Alberta's oil and gas industry where our ability to produce good, wholesome food is at risk of being compromised by the widespread, virtually unregulated use of this dangerous process."[629]

In Pennsylvania, 17,000 acres have already been lost to the development of natural gas fracking.[630] That land is not likely to be productive for several years because of "compaction and landscape reshaping," according to a study by the Penn State Extension Office.[631] U.S. Geological Survey scientists conclude there is a "low probability that the disturbed land will revert back to a natural state in the near future."[632]

The presence of natural gas drilling companies has also led to decreased milk and cheese production. Penn State researchers

Riley Adams and Dr. Timothy Kelsey concluded:

> "Changes in dairy cow numbers also seem to be associated with the level of Marcellus shale drilling activity. Counties with 150 or more Marcellus shale wells on average experienced an 18.7 percent decrease in dairy cows, compared to only a 1.2 percent average decrease in counties with no Marcellus wells. In contrast, the average county experienced a 6.4 percent decline in cow numbers. . . .
>
> "Higher drilling activity in all counties was associated with larger average declines in cow numbers. For example, counties with fewer than 5,000 cows in 2007 and no Marcellus wells averaged a loss of 2.2 percent, compared to an average 19 percent decline in such counties with 150 or more wells. Counties with 10,000 or more cows in 2007 and no Marcellus wells experienced an average 2.7 percent increase in cow numbers between 2007 and 2010, compared to an average loss of 16.3 percent in such counties with 150 or more Marcellus wells."[633]

Among other direct effects of fracking, according to Penn State agriculture specialists Dr. Patrick Drohan, Gary Sheppard, and Mark Madden, are problems attracting and keeping farm help, because of the higher pay from the fracking companies, and difficulty in obtaining mulch:

> [M]ulch is in high demand for erosion and sedimentation control on gas sites. As local supplies are exhausted, farmers who purchase mulch for animal bedding might expect their costs to increase. Gas companies in northeast Pennsylvania are importing mulch from Delaware, Maryland, Ohio, and Virginia.[634]

Destroying the Wildlife and Land

Significant impact upon wildlife is noted in a 900-page Environmental Impact Statement (EIS) conducted by the New York Department of Environmental Conservation, filed in September 2011. According to the EIS, "In addition to loss of habitat, other potential direct impacts on wildlife from drilling in the Marcellus Shale include increased mortality . . . altered microclimates, and increased traffic, noise, lighting, and well flares,"[635] also known as flaring or burn-offs. All natural gas companies use burn-offs to get rid of excess gas; this causes methane and other gases to enter the atmosphere. Flaring can produce as much as two million tons of carbon dioxide a year, according to a *New York Times* compilation of data.[636] The impact, according to the New York State report, "may include a loss of genetic diversity, species isolation, population declines ... increased predation, and an increase of invasive species."[637] The report concludes that because of fracking, there is "little to no place in the study areas where wildlife would not be impacted, [leading to] serious cascading ecological consequences."

A research study by the U.S. Geological Survey a year later, focusing upon two areas of Pennsylvania, reached the same conclusion. According to that report:

"Landscape disturbance associated with shale-gas development infrastructure directly alters habitat through loss, fragmentation, and edge effects, which in turn alters the flora and fauna dependent on that habitat. The fragmentation of habitat is expected to amplify the problem of total habitat area reduction for wildlife species, as well as contribute towards habitat degradation. . . .

"Changes in land use and land cover affect the ability of ecosystems to provide essential ecological goods and services, which, in turn, affect the economic, public health, and social benefits that these ecosystems provide."[638]

Also affected are humans who may eat fish that once lived in polluted streams and rivers, and cattle and other livestock that ate grass from polluted fields, drank polluted water, or absorbed airborne chemicals that were not detected during USDA inspections. Even vegetarians need to fear the effects of fracking. Toxic compounds "accumulate in the fat and are excreted into milk," says Dr. Motoko Mukai, environmental toxicologist at Cornell University.[639]

Every well "will generate a sediment discharge of approximately eight tons per year into local waterways, threatening federally endangered mollusks and other aquatic organisms," says Dr. Ronald Bishop, professor of biochemistry at the State University of New York, Oneonta.[640]

Research "strongly implicates exposure to gas drilling operations in serious health effects on humans, companion animals, livestock, horses, and wildlife," according to Dr. Michelle Bamberger, a veterinarian, and Dr. Robert E. Oswald, a biochemist and professor of molecular medicine at Cornell University. Their study, published in *New Solutions*, an academic journal in environmental health, documents evidence of milk contamination, breeding problems, and cow mortality in areas near fracking operations as higher than in areas where no fracking occurred. Drs. Bamberger and Oswald noted that some of the symptoms present in humans from what may be polluted water from fracking operations include rashes, headaches, dizziness, vomiting, and severe irritation of the eyes, nose, and throat. For animals, the symptoms often led to reproductive problems and death. At one farm, they documented 17 cows that died within an hour of being exposed to fracking fluids. Of the seven farms they studied in detail, "50 percent of the herd, on average, was affected by death and failure of survivors to breed."[641] Energy in Depth (EID)—the propaganda and disinformation organization formed and funded by energy companies,[642] and whose mission included opposing "new environmental regulations, especially with regard to hydraulic fracturing,"[643]— attacked the study as unscientific and "laughable at best, and dangerous for public debate at worst."[644] EID claimed the use of anonymous sources negated the study. However, Drs. Bamberger and Oswald followed acceptable scientific protocol; there are several reasons why there have not been additional studies or why sources aren't specifically named.

Elizabeth Royte, a distinguished science/environment writer, explains the problem:

"Rural vets won't speak up for fear of retaliation. And farmers aren't talking for myriad reasons: some receive royalty checks from the energy companies (either by choice or because the previous landowner leased their farm's mineral rights); some have signed nondisclosure agreements after receiving a financial settlement; and some are in active litigation. Some farmers fear retribution from community members with leases; others don't want to fall afoul of "food disparagement" laws or get sued by an oil company for defamation (as happened with one Texan after video of his flame-spouting garden hose was posted on the Internet. The oil company won; the homeowner is appealing).

"And many would simply rather not know what's going on. 'It takes a long time to build up a herd's reputation,' says rancher Dennis Bauste, of Trenton Lake, North Dakota. 'I'm gonna sell my calves, and I don't want them to be labeled as tainted. Besides, I wouldn't know what to test for. Until there's a big wipeout, a major problem, we're not gonna hear much about this.' Ceylon Feiring, an area vet, concurs. 'We're just waiting for a wreck to happen with someone's cattle,' she says. 'Otherwise, it's just one-offs'—a sick cow here and a dead goat there, easy for regulators, vets and even farmers to shrug off."[645]

In October 2006, an explosion at a gas compressor station led to the end of a successful horse breeding and boarding farm near DISH, Texas. According to an in-depth investigation by Peter Gorman for the *Fort Worth Weekly*:

"[Lloyd] Burgess, who had been out of town, returned to discover that one of his mares had aborted her foal. Two weeks later, the same thing happened to a second mare. . . .

"Several months later one of his stallions got sick and finally had to be put down. Then a mare went blind. Then another stallion, a valuable quarter horse, got sick and was saved only when a friend offered to take if off Burgess' property, away from the compressor stations on Burgess' back fence line, to nurse it back to health. . . .

" 'After the explosion and what happened to my horses, all my boarders took their horses out of there,' said Burgess. . . . 'Who could blame them? This was going to be my retirement, but now it's valueless.' "[646]

It probably also wasn't just a coincidence that vegetation died in the areas where fracking occurred. Gorman reported:

"[Burgess's] fence used to be lined by huckleberry trees, planted as a windbreak back in the 1930s and '40s. The wind blows through pretty freely now, however, since most of the trees have recently died. . . .

"[Jim] Caplinger said that when he moved into DISH, his street 'was lined with willows, and now they're almost all gone. They just died. And I've lost three elm and hackberry trees as well.' . . .

"[City Commissioner Bill Cisco] agreed that something very

troubling is happening to the trees. 'We've lost a lot of trees in DISH. And a number of them were cedar trees, which are almost impossible to kill. Those trees breathe just like we do, so when they start dying, you've got to pay attention,' he said. 'They're the canary in the coal mine for our air.' Thus far, dozens of trees in the little town and right outside it have died."[647]

DISH, in the middle of the gas-rich Barnett Shale and 15 miles from where the nation's first natural gas well was fracked, by the middle of 2012 had 60 wells and 11 compressor stations with 36-inch diameter pipes to transport the gas.[648]

In September 2009, about 65,000 fish, mussels and mudpuppies died in a 30 mile stretch of Dunkard Creek that flows in southwestern Pennsylvania and northern West Virginia.[649] It was one of the nation's largest fish kills. The cause was determined to be the presence of massive amounts of golden algae, which produce toxins that are lethal to certain aquatic life. The golden algae are found only in coastal areas, primarily the Gulf and southern states.[650] Extensive research for an environmental mystery determined that the algae could flourish in areas where there was fracking because of the extremely high salt content brought up in flowback water. The Pennsylvania Fish and Boat Commission charged Consol Energy with "illegal, toxic discharges [that were] willful, wanton and malicious."[651] Consol, which the Commission said was dumping flowback into Morris Run that flows into Dunkard Creek, denied the charges but agreed to pay $5.5 million to the U.S. Department of Justice and the West Virginia Department of Environmental Protection to settle civil claims and create a new water treatment plant for its operations.[652]

Wyoming's mule deer population declined by half during the past decade, possibly because of the addition of natural gas wells, according to wildlife biologist Dr. Hall Sawyer and biometrician Ryan Nielson. The authors noted: "Following the 2008 record of decision [on drilling], the level of winter drilling activity increased on the Mesa. It is possible that this increased winter disturbance affected fawn survival or adult reproduction."[653] Other reports about the effects of drilling upon other species are planned by the Bureau of Land Management.

Seven families were evacuated near Towanda, Pa., in April 2011, after about 10,000 gallons of wastewater contaminated an agricultural field and a stream that flows into the Susquehanna River, the result of an equipment failure.[654]

Wastewater accumulating in puddles was probably the cause of the death of a dog and horse in southwestern Pennsylvania, according to a *ProPublica* investigation. A veterinarian concluded, "The dog's organs began to crystallize, and ultimately failed" because of ethylene glycol present in the wastewater.[655]

Seventeen cows died after drinking contaminated water in a pasture near a natural gas rig in Caddo Parish, La., in April 2010. Neither the owner nor the subcontractors reported the spill; only after residents called the sheriff's department to report that cows were foaming at the mouth and bleeding that HazMat teams were dispatched, according to the *Shreveport Times*. Chesapeake Energy, which owned the well, acknowledged, "During a routine well stimulation/formation fracturing operation by Schlumberger for Chesapeake, it was observed that a portion of mixed 'frac' fluids, composed of over 99 percent freshwater, leaked from vessels and/or piping onto the well pad."[656] Chesapeake, of course, didn't state that it wasn't the 99 percent freshwater that caused the deaths but the one percent of whatever was in the toxic mix. Energy in Depth continually pushes the fiction about how pure the mixture is:

> "Did you know that 99.51% of hydraulic fracturing fluids are made up of sand and water? The last 0.49% is made up of household items like sodium chloride (table salt), guar gum (used in ice cream and baked goods) and citric acid (lemon juice)."[657]

Nevertheless, despite the fiction of the purity of the fracking solutions, Chesapeake and Schlumberger were each fined a token $2,000 by the Louisiana Department of Environmental Quality.[658]

One month after the deaths in Louisiana, the Pennsylvania Department of Agriculture quarantined 28 beef cattle on a farm in Tioga County following a spill from an impoundment pit that had stored wastewater. The spill affected about 1,200 square feet of pasture in an area where the cattle normally grazed. East Resources, which was mining the land, told farmers Carol and Don Johnson not to drink the well water, but didn't provide fresh water. East Resources later denied any correlation between the eventual death of several of the cattle and the fracking operation.[659] Even if there was no direct correlation, the mining operation caused problems for the Johnsons.

The installation of transmission lines across their property "spoiled almost every hay field I have," Don Johnson told reporter Chris Torres of *Lancaster Farming*.[660] In addition to having to deal with a loss of hayfields and toxic water spills, the Johnsons told the magazine they also had to deal with the noise from hundreds of trucks a week that use the couple's driveway to reach the well.

East Resources later told the Johnsons it wanted to put a second well and a compressor station onto property the Johnsons own a few miles away. But, Carol Johnson told the company she "had enough," and wanted to be left alone. "You sign a lot of stuff before you learn what's going on," she told *Lancaster Farming*.[661] The money they would earn from having leased mineral rights could help them survive the recession, but she says it would never be enough to compensate for the problems that the natural gas company created.

Joe Bezjak, of Smithfield, Pa., learned that same lesson. In his case, not only did a natural gas gathering company ruin part of his field, but a county judge put him in jail four days for protesting the pollution of his land.

Bezjak, 73, says he "hadn't even had a parking ticket" in his life and now, in December 2012, he was in jail.

His story begins in 2005, a year after he retired after four decades as a teacher and principal, when he allowed Atlas Pipeline Partners to lease almost 700 acres of his land that he, his wife, Mickey, and relatives owned in southwestern Pennsylvania. On that land, Bezjak and his family raised about 200 head of black Angus cattle. Atlas paid him $10 an acre. Years later, as farmers became more sophisticated in lease management options, the companies paid as much as $1,700 an acre. But this was still before the fracking boom, and Atlas paid only what it had to pay to get the rights to drill and put in pipes. And so, it put in a six-inch diameter transmission pipe to transport natural gas from a nearby well to a compressor station. Five years later, before the lease expired, the company created several shallow wells to assure continued mineral rights ownership. For seven years, the Bezjak family, cattle, and the natural gas company co-existed.

"There weren't any problems at first," Bezjak says, but Laurel Mountain Midstream, a joint venture of Williams Companies and Atlas Pipeline Partners, seven years later wanted to install a 16-inch line. Bezjak opposed it, but learned the company had the right to put in as many lines as it wanted in its right-of-way.

Laurel Mountain Midstream came onto his property in April 2012, without his permission, and tore down parts of his fence that divided the property. Bezjak had been careful to separate his herd by age and sex to maintain the quality of his herds. Without the fence, the cattle had wandered loose and interbred. "That ruined the herd," he says. "More than 40 years of careful work to separate the lines was destroyed."

The company was "all friendly at first and said they would put in the fence and make it right," says Bezjak. But it never was right.

In May, he confronted the workers about not properly restoring his fence. Someone from Laurel Mountain Midstream called the police, complaining that Bezjak threatened them with a .22 single-shot rifle. "I rode up to them on my Quad, and I had a rifle mounted on it; it's always been there, but I didn't use it or threaten anyone with it," he says, quickly pointing out he never used it—"I'd have trouble shooting anything." On advice of his State Police, he took the rifle off the Quad.

He'd walk onto his land, sometimes to see what the company was doing, sometimes to talk with the workers. One State Trooper told him, "Every time they see you, they call us." But still there were no heated arguments. During the Summer, one judge of the Common Pleas Court ordered the company to fix the fence to Bezjak's satisfaction; another judge ordered him not to talk to Laurel Mountain workers.

On Nov. 9, he saw workers pumping what he believed to be wastewater from a ditch onto his property. He says he told the workers to stop. "All farmers are stewards of the land," he says, "and I'm an environmentalist. I didn't like what they were doing." Neither did the state Department of Environmental Protection, which issued Laurel Mountain a notice of violation for discharging industrial waste into the waters of the commonwealth.[662] The DEP had previously cited Laurel Mountain in June for discharging pollution into public water[663] and in August not only for discharging industrial waste but also for failing to properly store, transport, process, or dispose of residual waste.[664] Laurel Mountain Midstream also failed to obtain a DEP precertification to lay pipelines. "Our concern is when pipelines crosses a stream or waterway, we need to know what they are going to do with the pipe" to avoid polluting the waters, says John Poister, community relations coordinator for the DEP Southwestern region.

On Nov. 28, Laurel Mountain workers used a backhoe to pile mud and dirt on contaminated water. Bezjak called the DEP, whose inspectors in the region quickly respond to problems, but was told that there were not enough personnel available to investigate that day. But, the DEP did tell him that that covering up a spill was not acceptable, and it would investigate. So, Bezjak "decided to confront them myself to stop it." He says, he "was concerned they were destroying the environment, and told them so." At the time, he believed the polluted water had come from a terra cotta pipe that had carried sulfur water from surface mining decades earlier. However, the water could also have come from a pipe laid down by Laurel Mountain. "We just don't know," says Poister.

Whatever its origins, several hundred gallons of acidic water drained onto parts of the land. "I had been farming that land 40 years," Bezjak says, "and had developed a good pH value and good grass." And now, not only was contaminated water lying in sulfur-red pools on his land, but heavy rains were washing it into the wetlands and into a small stream. The DEP had previously measured the water to have a pH reading of 3.36, acidic and dangerous. A private company Bezjak hired confirmed the water as toxic.

Bezjak readily acknowledges, "I got upset and was arguing with them because I saw what I'd worked for my whole life being damaged by these people who lied to me. So I stopped them, and by 10 a.m. there were three State Police troopers in my driveway and I was handcuffed." He was not arrested. "I wasn't looking for trouble," says Bezjak, "but they were damaging the land." A supervisor, says Bezjak, "even apologized to me and said the company would 'make it right.'"

Two weeks later, Bezjak got a summons. In Court four days after he received the summons, Common Pleas Court Judge Nancy D. Vernon "was livid that I had defied a court order not to talk to the workers." She didn't want to hear about the supervisor's apology, nor did she allow Bezjak to present evidence about environmental pollution. What upset her, he says, "is that I violated her order not to talk with the workers, and then in court told her she had to do whatever she had to do." He had planned to face the charge, pay whatever fine the judge imposed, continue Christmas shopping, and then pick up a friend he had taken for chemotherapy. "I never expected to go to jail over this," he says.

Citing Bezjak for contempt, Vernon ordered him handcuffed and jailed without bail in the Fayette County Prison for four days, beginning Dec. 14.

On the Friday afternoon that Vernon threw him in jail, Laurel Mountain Midstream, by DEP direction, came onto his property to backfill the trench to temporarily stabilize the site during Winter. "We were concerned about acid water seepage, and wanted to minimize the potential of accelerated erosion," says John Poister. DEP, says Poister, planned to require Laurel Mountain Midstream to fully restore the site during the Spring thaw.

In a prepared statement, Julie Gentz said her company's goal is "to have good working relationships with our landowners and other stakeholders. This situation is a rare occurrence, but there have been a number of disagreements with this particular landowner that we have tried to resolve amicably."[665]

Like many farmers in the Marcellus Shale, Bezjak doesn't oppose drilling for natural gas. What he does oppose is polluting the public health and environment. And, like most farmers, Bezjak's experience is that the landmen for companies involved in fracking "start out trying to be your friend, but they just lie and lie and lie to you."

He now keeps his cattle out of that area, "because I don't know what's in that water, and I don't want to contaminate the herd or cause health problems for them or the public."

PHOTO: Gary F. Clark

Trucks and equipment at a rig near Warrendale, Pa.

CHAPTER 11
Wastewater Removal

Horizontal hydraulic fracturing requires not only the injection of millions of gallons of water, sand, and chemicals, many of them toxic, into the earth, but also their removal. More than 30 trillion gallons of toxic waste have been injected into almost 700,000 underground waste and injection wells in 32 states over the past few decades, according to a data analysis by *ProPublica*.[666] In 2011, natural gas companies drilling in Pennsylvania had to get rid of 31 million gallons of wastewater. Since 2002 when the first two permits were issued, to the beginning of 2013, energy companies that used fracking to extract natural gas in Pennsylvania had to dispose of about 1.15 billion gallons of wastewater.[667]

Among toxic chemicals in wastewater are benzene, toluene, arsenic, and nitrates. Although nitrates in groundwater can come from several sources, they are also in the wastewater from fracking. Wastewater, in addition to bringing up several elements, contains radioactive materials, including Uranium-238, which decays into Radon, one of the most radioactive and toxic gases. Radon is the second highest cause of lung cancer, after cigarettes, according to the EPA; for non-smokers who develop lung cancer, radon exposure is the leading cause of death.[668] Like carbon monoxide, it is odorless, colorless, and detectable only with sophisticated measuring instruments.

Analysis of wastewater from 13 sites in New York state revealed the presence of Radium-226, which has a half-life of 1,601 years, and can lead to bone cancer, leukemia, and lymphoma.[669] Some of the radium was more than 250 times the acceptable range for discharge into the environment, according to research conducted by Dr. Marvin Resnikoff.[670] A U.S. Geological Survey of well samples collected in Pennsylvania and

New York between 2009 and 2011 revealed that 37 of the 52 samples had Radium-226 and Radium-228 levels that were 242 times higher than the standard for drinking water. One sample, from Tioga County, Pa., was 3,609 times the federal standard for safe drinking water, and 300 times the federal industrial standard.[671]

Radium in wastewater brine was three times higher from fracking than from conventional wells, according to the U.S. Geological Survey.[672] The *Columbus* (Ohio) *Dispatch* reported that in 37 of the 52 samples the USGS analyzed, "radioactivity from Radium-226 and Radium-228 was at least 242 times higher than the drinking water standard and at least 20 times higher than the industrial standard."[673]

An investigation by *New York Times* reporter Ian Urbina, based upon thousands of unreported EPA documents and a confidential study by the natural gas industry, concluded, "Radioactivity in drilling waste cannot be fully diluted in rivers and other waterways." Urbina learned that wastewater from fracking operations was about 100 times more toxic than federal drinking water standards; 15 wells had readings about 1,000 times higher than standards.[674] Urbina's article, part of the "Drilling Down" series, was attacked by conservative politicians and talk show pundits, and by the natural gas industry. "Drilling Down [is] a series of articles with ominous headlines and dubious facts," Energy for America argued.[675] However, because Urbina documented his sources and verified the facts, the *Times* didn't have to print corrections.

No matter what the public relations practitioners of the natural gas industry believe, or want citizens to believe, there are five methods to remove wastewater, all of which "present significant risks of harm to public health or the environment," according to Rebecca Hammer, attorney with the National Resources Defense Council, and Dr. Jeanne VanBriesen, professor of civil and environmental engineering at Carnegie-Mellon University, authors of the NRDC's 113-page study, *In Fracking's Wake* (May 2012).[676]

Use of Wastewater as a Deicer On Roads

The salinity of wastewater makes it an ideal deicer. In some cases, wastewater is used to reduce dust on dirt roads.

However, because of the toxicity, wastewater will be absorbed into the roads or washed into nearby streams, creeks, and rivers, further poisoning the environment. Salts spread on roads and agriculture fields could contain ammonia, arsenic, diesel hydrocarbons, lead, mercury, and various volatile organic compounds.[677] Certain forest animals will also be tempted to lick the salt-laden wastewater that includes not just the toxins but also radioactive elements.

Until October 2012, the practice was forbidden in Pennsylvania. However, the DEP granted Integrated Water Technologies a 10-year general permit to spread chemical salts from the Marcellus Shale wastewater on all public roads and fields in the state.[678]

In an appeal filed with the Environmental Hearing Board, Citizens for Pennsylvania's Future (PennFuture) argued that the original permit allowed the company only to process wastewater, and charged that DEP "misled the public about the nature and scope of the permit" to now allow "beneficial use." The DEP action, said PennFuture, was "arbitrary, unreasonable, and contrary to law" by deliberately broadening that permit without a required public hearing. The DEP response called the appeal "baseless [and] an attempt to manufacture a controversy."[679]

However, the appeal was definitely not baseless. "We need to know what the company is claiming it can remove from the fracking flowback wastewater," said Dr. John Stolz, director of the Center for Environmental Research and Education at Duquesne University.[680] The problem, said Dr. Stolz, "is that flowback chemical composition can vary widely depending on how long the fluids have been in the [well]. It becomes saltier the longer it's been underground and the quality of the water itself can be very different."[681]

Storage in Open Pits

Wastewater stored in open pits also evaporates, causing air pollution. A report by the Citizens Marcellus Shale Commission in October 2011 pointed out that Pennsylvania has "no requirement to store [fluids and wastewater] in a closed loop system," and some companies will mix the wastewater with freshwater. "Drilling wastes frequently are stored in pits with synthetic liners, which often have leaked and caused pollution," according to the Commission, which conducted public hearings throughout Pennsylvania.[682] The Commission was composed of representatives from major health, environmental, and community organizations; it disbanded after issuing its final report.

Additional problems occur when gas companies deliberately disregard health and environmental issues.

"In the beginning we would dig a hole and then we'd just throw plastic in it. That was more or less to make the homeowners feel comfortable about us drilling on their property," Scott Ely, a former employee of GasSearch Drilling, a subsidiary of Cabot Oil and Gas, told Laura Legere of the Scranton (Pa.) *Times-Tribune.* Ely said he once watched Cabot employees push a "big, goopy concoction" of sand, gels, and fracking fluid acids over a bank. The oil and gas industry "had no care for what spilled anywhere. It was the most reckless industry I've ever seen in my life," said Ely.[683]

Rules issued by the U.S. Bureau of Land Management in May 2012 require companies to store wastewater in closed tanks or above ground pits that have impervious plastic liners.[684] However, in keeping with the industry-wide mind-set, Kathleen Sgamma of Western Energy Alliance, a trade association, said that regulations "add more delays and cost onto an already excessively bureaucratic federal process."[685]

Recycling

Many natural gas drillers use wastewater as a primary source of water to fracture the shale. In Pennsylvania, about one-third of all wastewater is recycled.[686] "On-site recycling can have significant cost and environmental benefits as operators reduce their freshwater consumption and decrease the amount of wastewater destined for disposal," say the NRDC's Rebecca Hammer and Dr. Jeanne VanBriesen. However, they conclude, recycling "can generate concentrated residual by-products (which must be properly managed) and can be energy-intensive."[687] When used as a primary fluid in hydraulic fracturing, recycled wastewater is not subject to the Safe Water Drinking Act.

Sewage Treatment

Because of the nature of the Marcellus Shale deposit in Pennsylvania, as opposed to neighboring states, natural gas companies transport most of the wastewater to other states for reuse or disposal or take it to sewage treatment plants. About half of all Pennsylvania wastewater was taken to treatment plants in 2011. The plants discharge the treated wastewater into the state's rivers. However, present methods can't remove all the salt and some other chemicals and radioactive elements. "You are asking sewage plants to do something that they were not designed to do," says Robert F. Kennedy Jr.[688]

Total Dissolved Solids (TDS) in wastewater brought up from more than a mile below the earth's surface may have as much as five times the salinity as ocean water.[689]

TDS was responsible for machinery corrosion at several industrial plants, including U.S. Steel and Allegheny Energy, along the Monongahela River near Pittsburgh. Tests by the Department of Environmental Protection in October 2008 measured TDS contamination at twice the maximum acceptable levels.[690] However, Stephen W. Rhoads, president of the Pennsylvania Oil and Gas Association in 2009, claimed most of the TDS came not from wastewater but from abandoned mines.[691]

In April 2011, the DEP finally ordered the industry to cease deliveries to the 15 facilities that had currently taken wastewater.[692] The following month, the EPA urged the DEP to take stricter measures in handing wastewater.[693]

Underground Injection

More than two billion gallons of drilling waste are poured into injection wells every day.[694] According to the EPA:

"Underground injection has been and continues to be a viable technique for subsurface storage and disposal of fluids when properly done. ...EPA recognizes that more can be done to enhance drinking water safeguards and, along with states and tribes, will work to improve the efficiency of the underground injection control program."[695]

However, a *ProPublica* investigation by Abrahm Lustgarten revealed:

"Records from disparate corners of the United States show that wells drilled to bury this waste deep beneath the ground have repeatedly leaked, sending dangerous chemicals and waste gurgling to the surface or, on occasion, seeping into shallow aquifers that store a significant portion of the nation's drinking water."[696]

Between 2007 and 2010 more than 17,000 violations were issued against companies for problems associated with injection wells; more than 220,000 well inspections revealed "structural failures inside wells are routine," Lustgarten reported.[697]

Pennsylvania has only five active deep injection wells.[698] Because of the nature of the Marcellus Shale, underground injection in most areas of the state isn't possible, so Pennsylvania drillers transport wastewater to Ohio. In Ohio, about 12.2 million barrels of wastewater, 53 percent of it trucked in from Pennsylvania and West Virginia—including 90 percent of all of Pennsylvania's waste water—were injected into underground wells in 2011, according to an investigation by Rick Reitzel of WCMH-TV, Columbus.[699]

On Jan. 17, 2001, gas collected in an underground injection well near Hutchinson, Kansas, ignited, destroyed two buildings, damaged 25 others, killed two residents, and forced hundreds to evacuate. Abrahm Lustgarten observed:

"Among a small community of geologists and regulators . . . the explosions in Hutchinson—which ranked among the worst injection-related accidents in history—exposed fundamental risks of underground leakage and prompted fresh doubts about the geological science of injection itself.

"Geologists in Hutchinson determined that the eruptions had sprung from an underground gas storage field seven miles away. For years, a local utility had injected natural gas between 600 and 900 feet down into old salt caverns, storing it in a rock layer believed to be airtight so that it could later be pumped back out and sold. The gas had leaked out and migrated miles into abandoned injection wells once used to mine salt, then shot to the surface.

"'It was an unusual event,' said Bill Bryson, a member of the Kansas Geological Survey and a former head of the Kansas Corporation Commission's oil and gas conservation division. 'Nobody really had a feeling that if there was a leak, it would travel seven miles and hit wells that were unknown.'

"Though regulated under different laws than waste injection wells, gas storage wells operate under similar principles and assumptions: that deeply buried layers of rock will prevent injected substances from leaking into water supplies or back to the surface.

"In this case the injected material had done everything that scientists usually describe as impossible: It migrated over a large distance, travelled upward through rock, reached the open air and then blew up.

"The case, described as 'a continuing series of geologic surprises and unexpected complexities' by the Kansas Geological Survey, flummoxed some of the leading injection experts in the world.

"Perhaps more troubling was that some of the officials assumed to be most knowledgeable about injection wells and the risks of underground storage seemed oblivious to the conditions that led to the accident."[700]

ONEOK Inc., which identifies itself as "among the largest natural gas distributors" in the U.S., later acknowledged there was a leak of 143 billion cubic feet of gas.[701]

EARTHQUAKES

Explosions are only one problem associated with injection wells. The Railroad Commission of Texas claims not only is fracking "an environmentally safe process," but that "the Commission has no data that links hydraulic fracturing activities to earthquakes."[702] Scientists disagree.

"We've known for decades that injection of fluids can and does trigger earthquakes in some cases," Dr. John Cassidy, a seismologist with Natural Resources Canada, told the *Toronto Star*.[703]

"Pumping water into the ground is one of the most mechanically destructive things that we can do," says Dr. Richard Ketcham, associate professor of geological sciences at the University of Texas.[704]

The first sentence of a 240-page research report by the National Research Council (NRC), working with the U.S. Department of Energy, is: "Since the 1920s we have recognized that pumping fluids into or out of the Earth has the potential to cause seismic events that can be felt."[705]

The report, published in June 2012, pointed out:

"Although only a very small fraction of injection and extraction activities at hundreds of thousands of energy development sites in the United States have induced seismicity at levels that are noticeable to the public, seismic events caused by or likely related to energy development have been measured and felt in Alabama, Arkansas, California, Colorado, Illinois, Louisiana, Mississippi, Nebraska, Nevada, New Mexico, Ohio, Oklahoma, and Texas."[706]

NRC research suggested that deep injection wells themselves, rather than the actual fracking process, was the problem.

Dr. Arthur McGarr, a geophysicist with the U.S. Geological Survey, developed a formula based upon field research that reveals the volume of fluid and solids put into the earth is proportional to the size of any subsequent earthquakes. He says there is no way to determine when or if earthquakes will occur, only that when they do occur, the formula of volume-to-earthquake size applies.[707]

"If we have more wells, we have more chance of events [and] if we have more events, there's more probability of higher magnitude events," Dr. Murray Hitzman told CNN.[708] Dr. Hitzman is professor of geology at the Colorado School of Mines and chair of the committee formed by the National Research Council to study earthquake probabilities from fracking.

An investigation by U.S. Geological Survey (USGS) scientists suggests that earthquakes near oil and gas drilling sites are "almost certainly man-made." Research data presented to the April 2012 meeting of the Seismological Society of America suggests that high-pressure forcing of wastewater into injection wells can cause low-level earthquakes. According to the six-member team, headed by Dr. William L. Ellsworth:

"A remarkable increase in the rate of [magnitude-3.0] and greater earthquakes is currently in progress. . . . A naturally-occurring rate change of this magnitude is unprecedented outside of volcanic settings or in the absence of a main shock."[709]

Three USGS scientists who were part of the study presented a companion report in December 2012 to the American Geophysical Union. An earthquake near Trinidad, Colo., that registered a magnitude level of 5.3 in August 2011 "renewed interest in the possibility that an earthquake sequence in this region that began in August 2001 is the result of industrial activities," according to the team. Focusing upon earthquakes in the Raton Basin of northern New Mexico and southern Colorado, they concluded:

"the majority, if not all of the earthquakes since August 2001 have been triggered by the deep injection of wastewater related to the production of natural gas from the coal-bed methane field here. The evidence that this earthquake sequence was triggered by wastewater injection is threefold. First, there was a marked increase in seismicity shortly after major fluid injection began in the Raton Basin. From 1970 through July of 2001, there were five earthquakes of magnitude 3 and larger located in the Raton Basin. In the subsequent 10 years from August of 2001 through the end of 2011, there were 95 earthquakes of magnitude 3 and larger. . . . Second, the vast majority of the seismicity is located close (within 5km) to active disposal wells in this region. . . . Finally, these wells have injected exceptionally high volumes of wastewater. The 23 August 2011 [magnitude 5.3] earthquake, located adjacent to two high-volume disposal wells, is the largest earthquake to date for which there is compelling evidence of triggering by fluid injection activities; indeed, these two nearly-co-located wells injected about 4.9 million cubic meters of wastewater during the period leading up to the M5.3 earthquake, more than 7 times as much as the disposal well at the Rocky Mountain Arsenal that caused damaging earthquakes in the Denver, CO, region in the 1960s."[710]

Cuadrilla Resources acknowledged its use of hydraulic fracturing was the cause of two small earthquakes near Lancashire, England, in 2011.[711] A 2.3 magnitude quake in April, and a 1.4 quake the following month led Cuadrilla to suspend operations and initiate an independent report.

An investigation conducted by the British Columbia Gas and Oil Commission revealed that 272 small earthquakes in the

Horn River Basin may have been caused by fracking between April 2009 and December 2011. Of the 272, the Commission determined that 38 measured between 2.2 and 3.8 magnitude on the Richter scale; the others were less than 2.2 magnitude. No earthquakes were detected prior to the beginning of natural gas fracking, according to the Commission. The Commission called for "pre-emptive steps to ensure future events are detected and the regulatory framework adequately provides for the monitoring, reporting and mitigation of all seismicity related to hydraulic fracturing, thereby ensuring the continued safe and environmentally responsible development of shale gas within British Columbia."[712]

Within a week in March 2010 five small earthquakes, the largest measuring 2.8 on the Richter Scale, were detected near Cleburne, Texas, a Dallas suburb. In 2011, four earthquakes hit the area; fourteen earthquakes in June and July 2012, each one between 2.3 and 3.5 on the Richter scale, were recorded.[713]

"If you look at the history of earthquakes in Texas," says geophysicist Paul Caruso, "it's unusual to have any down there."[714] Also unusual were earthquakes in Timpson, Texas, sitting in the middle of the Haynesville Shale formation about 50 miles east of Cleburne. In May 2012, two earthquakes, with their epicenters six miles apart, struck Timpson. The first one registered 3.9 on the Richter scale; the second one, a week later, registered 4.3. "I've had people who've lived here all their lives who told me they've never experienced anything like this," said Larry Burns, Timson's emergency response coordinator.[715] Within a 10 mile radius of Timpson are seven to ten waste injection wells; within Shelby County are more than 540 natural gas wells. Burns said he suspected pressure in injection wells was the cause for the earthquakes, and gas field workers in the area had told him of a possible connection.[716]

Between 1972 and 2009, two to six earthquakes were recorded per year in Oklahoma.[717] In 2010, after natural gas companies had increased their operations, the state recorded 1,047 earthquakes; 103 of them could be felt by residents, according to the Oklahoma Geological Survey (OGS).[718] As of January 2011, Oklahoma had 43,600 active gas well and 10,500 injection/disposal wells.[719]

A series of about 50 earthquakes within two miles of a

190

drilling site in Garvin County, Okla., may have been caused by fracking operations. The earthquakes, all within a 24-hour period, beginning Jan. 18, 2011, may have been "induced by hydraulic fracturing," according to Austin Holland of the Oklahoma Geological Survey. Each earthquake measured between 1.0 and 2.8 on the Richter scale.[720]

Arkansas, in March 2011, banned the use of injection wells following a series of 700 small earthquakes within a six month period. The Arkansas Oil and Gas Commission issued the emergency temporary moratorium less than a week after the largest quake, measuring 4.7 on the Richter Scale, hit north-central Arkansas; that quake was the largest in the state in 35 years.[721] There are about 700 disposal wells in Arkansas. Although there was no direct evidence that injection well fracking caused the earthquakes, during the next three months after the moratorium there was a significant decrease in earthquakes, said Matt DeCample, assistant to Gov. Mike Beebe.[722] Arkansas issued a permanent moratorium in August.[723]

The fracking process probably caused a dozen earthquakes between March and December 2011 near Youngstown, Ohio, according to an investigation by the Ohio Department of Natural Resources (DNR),[724] which had initially claimed there was no correlation.[725] The quakes, all within one mile of Northstar 1, a 9,000 foot deep injection well, registered between 2.1 and 4.0 magnitude on the Richter scale. It is "virtually certain" that the fracking process caused a tremor on Dec. 31, 2011, measuring 2.7 on the Richter scale, Dr. John Armbruster, a seismologist at Columbia University told CNN.[726] An investigation by Mike Ludwig of *Truthout* suggests that the earthquakes may have been triggered after the state twice allowed D&L Energy to increase maximum pressure to 2,250 pounds per square inch.[727]

On Jan. 1, 2012, Gov. John Kasich, who strongly favors gas drilling, placed a moratorium on injection well drilling in the area around Northstar 1.[728] Two months later, the Ohio DNR issued a new set of regulations for natural gas drillers that use fracking. Among the tougher rules are those which prohibit brine injection into Precambrian rock formations and plugging existing injection sites, better monitoring of all injection fluids and wastewater, and the requirement that all drillers planning to use fracking must submit more comprehensive geological data when requesting permits.

"You can't prove that any one earthquake was caused by an injection well, but it's obvious that wells are enhancing the probability that earthquakes will occur," says Dr. Cliff Frohlich, senior research scientist at the University of Texas Institute for Geophysics.[729] Dr. Frohlich's research found not only a strong correlation between earthquakes and injection wells in the Barnett Shale in Texas, but "more of the smaller ones than were previously known."[730]

Rebecca Hammer and Dr. Jeanne VanBriesen conclude "[T]here are not sufficient rules in place to ensure any of [the wastewater disposal methods] will not harm the people or ecosystems."[731] Hammer and Dr. VanBriesen believe only recycling and underground injection methods, with adequate safety standards, could be used to deal with wastewater. The NRDC position is that the use of wastewater as deicer on roads, storage in open pits, and sewage treatment with discharge into rivers "present such great threats that they should be banned immediately."[732]

Platform on rig near Warrendale, Pa.

CHAPTER 12:
Government and Industry Response

The response by the industry and its political allies to the scientific studies of the health and environmental effects of fracking "too often has been to argue that hydraulic fracturing can't possibly cause any problems," says Fred Krupp, president of the Environmental Defense Fund.[733]

The energy companies claim water wells were tainted before gas operations began, "but only after fracking began did these chemicals show up in people's water supplies," says Dr. Amy Paré, a physician from McMurray, Pa. Hit hard by the public response to industry denials, and faced by numerous photos and videos showing contaminated water, the industry is now testing the water before and after fracking occurs.

"In general [the natural gas industry tends] to be dismissive of individual complaints while expressing an understandable need for further research and concern for the health of individuals, but really shying away from any connection with their own activities," Abrahm Lustgarten told NPR's "Fresh Air Report." Lustgarten says:

> "You won't hear the drilling industry say, 'This isn't an issue and we don't have to study this.' You won't hear them say they don't care about [individuals with health problems]. But you will hear them say [certain individuals appear] to not like the industry and maybe [they have] health issues and maybe [they want] the industry to leave. . . And by the way, we need to do a whole lot more research and it's a decade-long effort and let's just get started and not talk about blame at this point."[734]

The volume and intensity of response "has approached the issue in a manner similar to the tobacco industry that for many years rejected the link between smoking and cancer," say Drs. Michelle Bamberger and Robert E. Oswald.[735]

The response to a suit claiming health problems caused by a three million gallon wastewater spill was not to address the issue but to attack the attorney representing three families. Matt Pitzarella, director of corporate communications and public affairs for Range Resources–Appalachia, accused of the spill, said the suit "isn't about health and safety; it's unfortunately about a lawyer hoping to pad his pockets, while frightening a lot of people along the way."[736] The Haney, Kiskadden, and Voyles families of Washington County had sued the company after developing "a multitude of health problems, including nose bleeds, headaches and dizziness, skin rashes, stomach aches, ear infections, nausea, numbness in extremities, loss of sense of smell and bone pain," according to the Pittsburgh *Post-Gazette*.[737] However, a major part of the suit accuses Range, several sub-contractors and two water testing labs for not revealing full and accurate reports of the contamination, and for advising the families, as the *Post-Gazette* reported, to believe it was safe to "drink, cook, and bathe in the contaminated water." A separate lawsuit filed by Beth Voyles, in May 2011, had asked for a *writ of mandamus* to force the Pennsylvania DEP to fulfill its obligations to protect public health. That suit stated Voyles had "numerous health ailments, including but not limited to rashes, blisters, light-headedness, nose bleeds, lethargy, and medical testing [that] revealed elevated levels of arsenic, benzene, and toluene in her body."[738] Those chemicals are found in fracking fluids, which are alleged to have been present in a nearby impoundment that had a capacity of about 321,000 barrels of wastewater. "Even after numerous reports of holes and tears in the primary impoundment liner that needed to [be] repaired [were reported], the DEP still required no monitoring system," the *Canon-McMillan Patch* reported.[739]

POLITICAL INFRINGEMENT UPON SCIENCE

"Offensive" is how Sen. James Inhofe (R-Okla.) described the

EPA report that documented water pollution in Pavillion, Wyo., as being caused by the chemicals used in the fracking process. He claimed the EPA's study was an "apparent effort to reach a predetermined conclusion that hydraulic fracturing affects groundwater. According to Inhofe, the EPA's conclusions . . .

"if made in an irresponsible way, could have devastating effects on natural gas development as well as our economy. One study has shown that in 2010 the shale gas industry supported more than 600,000 jobs, and this number will grow to 1.6 million jobs by 2035. EPA's reckless process could put these jobs at risk.

"EPA has gotten off to a dubious start and going forward, its investigation can have no credibility if it is not held to the highest standards."[740]

Inhofe claimed, "It is irresponsible for EPA to release such an explosive announcement without objective peer review."[741] Inhofe and nine other Republican senators demanded the EPA classify the research as "highly influential scientific assessment," and subject it to a more rigorous "peer review" process. The senators' letter to the EPA followed a similar request by Encana, the wells' owner. However, the review process following a draft report includes peer review and public hearings, as was done by the EPA.

Inhofe's concern may have been more political than scientific. Inhofe, ranking member of the Senate's Environment and Public Works committee, received $655,500 in campaign contributions from individuals and PACs associated with the oil and gas industry during the 2011–2012 election cycle,[742]and $1.4 million since 1989,[743] according to data compiled by the Center for Responsive Politics.

Also attacking the EPA report were Wyoming politicians and newspapers, which echoed industry response. "The EPA may have poisoned the public debate by releasing its report," the *Casper Star-Tribune* editorialized.[744] The newspaper called the report "clumsy" and claimed the samples were "improperly tested." Disregarding obvious ties between the industry and politics, the newspaper attacked the researchers. The *Star-Tribune* claimed the EPA research was "not about science and more about politics." In an OpEd rebuttal, James B. Martin,

EPA regional administrator, stated the newspaper's claims "do a disservice to the rigorous scientific process EPA conducted," and pointed out the three year study was reviewed by EPA managers and "subjected to an initial peer review by independent experts."[745]

Two months after the EPA draft report was published, the House Subcommittee on Energy held a hearing to question the EPA methods. The Committee determined the EPA "wells were drilled and installed without the State of Wyoming's knowledge or assistance. Without these records, it is difficult to eliminate the possibility that EPA's actions in drilling and installing the monitoring wells may have contributed to the contamination detected in the samples."[746]

Michael Krancer, Pennsylvania's DEP secretary, had a broader concern than just the Pavillion study. The previous year, he had said, "The myth that terrible chemicals are getting into the groundwater is completely myth. It is bogus."[747] Now, before a Congressional hearing, Krancer claimed all studies that showed toxic methane gas in drinking water were "bogus," and specifically cited as "statistically and technically biased"[748] a Duke University study.[749] Two of the study's researchers fired back. In an OpEd article in the *Philadelphia Inquirer*, Drs. Robert Jackson and Avner Vengosh suggested, "Rather than working to discredit any science that challenges his views, the secretary and his agency should be working to get to the bottom of the science with an open mind."[750]

Representing the interests of the gas lobby, Karen Harbert of the U.S. Chamber of Commerce said she didn't want to see states enact restrictive financial and environmental laws.[751]

REPORTING THE TRUTH

Although the natural gas industry may lie or shade the truth to the media, the people, and the governing bodies, they are required to report the truth to the Securities and Exchange Commission. In its annual SEC 10-K report, filed December 2011, Cabot Oil & Gas Corp. admitted:

> "Our business involves a variety of operating risks, including:
> • well site blowouts, cratering and explosions;

• equipment failures;
• pipe or cement failures and casing collapses, which can release natural gas, oil, drilling fluids or hydraulic fracturing fluids;
• uncontrolled flows of natural gas, oil or well fluids;
• fires;
• formations with abnormal pressures;
• handling and disposal of materials, including drilling fluids and hydraulic fracturing fluids;
• pollution and other environmental risks; and
• natural disasters.

"Any of these events could result in injury or loss of human life, loss of hydrocarbons, significant damage to or destruction of property, environmental pollution, regulatory investigations and penalties, suspension or impairment of our operations and substantial losses to us.

"Our operation of natural gas gathering and pipeline systems also involves various risks, including the risk of explosions and environmental hazards caused by pipeline leaks and ruptures. The location of pipelines near populated areas, includ- ing residential areas, commercial business centers and Industrial sites, could increase these risks."[752]

The final word about the health of the public is not that of any of the scientists who have been researching the health and environmental issues surrounding the fracking process. The final word is "mitigation," and it belongs to Rex W. Tillerson, the CEO of ExxonMobil, which earned a $41 billion net profit in 2011[753] and $34.9 billion profit for the first three quarters of 2012.[754] In a speech to the Council on Foreign Relations in June 2012, the same speech that he declared people are illiterate in the sciences, and that opponents of the energy industries are manufacturing fear, Tillerson laid out a corporate truth:

"And so when people manufacture this fear that we can't allow this to go forward . . . our answer is 'yes we can,' because we will have a technological solution and we will have risk mitigation and risk management practices around those resources to ensure they can be developed in a way that mitigates risk—it doesn't eliminate it, but when you put it into the risk versus benefit balance, it comes back into a balance that most reasonable people in society would say, 'I can live with that.' "[755]

Thus, the energy industry is telling the people there will be accidents. There will be deaths. There will be health and environmental consequences. But, they are acceptable because "mitigation" allows a corporation to accept errors, injuries, illnesses, environmental destruction, and even death if they believe there is a "greater [financial] good" that outweighs those risks.

It is no different than when Ford in the 1970s manufactured the Pinto, a sub-compact with a gas tank positioned at the rear of the vehicle, and didn't put in a heavyweight bumper or much reinforcement between the rear panel and the gas tank. This design flaw led to a higher-than-expected number of fires if the car was hit from the rear. To recall each Pinto and modify the fuel tank would cost $11 per vehicle,[756] about $121 million total. But, Ford figured that cost would be greater than what it would have to pay out for injuries and deaths caused by the defect, which it figured to be about $50 million. Thus, "mitigation" and "risk management" suggested it would be cheaper to allow injuries and deaths than to fix the problem. A leaked memo published in *Mother Jones* magazine[757] led to public outrage, additional lawsuits, a government-mandated recall, and one of the worst public relations disasters in Ford's history.

In four decades, the energy industry learned nothing from the fall-out over Ford's decision to sacrifice lives to cost-ratio benefits.

PHOTO: Diane Siegmund

New pipes on Southside Road near Leroy, Pa.

CHAPTER 13:
The People Push Back

Unable to match the money spent for billboards, brochures, and media ads by the natural gas industry and its front groups, the anti-fracking movement is using social media and mass demonstrations to get its message out. Financial donations, not corporate sponsorships, help create the signs and communications. More than 500 blogs and websites, most of them organized by non-profit organizations or representing a small coalition of local and regional organizations, are spreading the message about problems caused by fracking.

More than 60 groups in Pennsylvania, including chapters of national environmental organizations, oppose fracking. Three of the most active groups of grassroots activists are the Gas Drilling Awareness Coalition (GDAS) of Luzerne County in northeastern Pennsylvania; Shale Justice, a coalition of several groups, that was created in January 2013 to develop a unified strategic message and marketing campaign to match the work of volunteer-citizens against corporate budgets and full-time staff; and Organizations United for the Environment of the Susquehanna Valley region in north-central Pennsylvania. OUE was founded in 1972 to unite environmental activists, and is one of the state's leaders in opposing both environmental destruction and fracking.

From small village meetings, with only one or two members willing to speak against the incursion of the wells, to mass rallies of thousands, protestors are active in promoting the message that fracking for natural gas can harm public health, animal life, and the environment.

More than 5,000 marched from the U.S. Capitol to the headquarters of the American Petroleum Institute, July 28, 2012. A month later, more than 1,000 individuals marched on Albany, N.Y., to target Gov. Andrew Cuomo who had given suggestions he might approve fracking for five southern tier counties. A newly-formed coalition, New Yorkers Against Fracking, had helped organize the protest. Leaders are environmental expert Dr. Sandra Steingraber; Lois Gibbs, who had organized the Love Canal Homeowners Association in 1978 and then created the Center for Health, Environment and Justice, which provided assistance for more than 11,000 grassroots groups; musician/singer Natalie Merchant, who has a long history of environmental and human rights activism; and actor Mark Ruffalo, who has spoken out on numerous radio and TV shows about the destruct-tion of the environment; Ruffalo has also written numerous columns and OpEds, and is a frequent speaker at rallies against fracking and for the environment.

More than 1,000 protested an industry-sponsored event at Philadelphia's Convention Center, Sept. 20, 2012, that mixed energy company executives and politicians. Two days later, Global Frackdown Day, the voices of the "fractivists" were heard in small towns and urban areas in the United States and overseas in more than 100 events organized by Food & Water Watch.

In late Summer 2012, Sean Lennon and Yoko Ono created Artists Against Fracking. Among almost 200 artists are Alec Baldwin, Jackson Brown, David Crosby, Robert DeNiro, Anthony Edwards, Jimmy Fallon, Carrie Fisher, Roberta Flack, David Geffen, Daryl Hanna, Anne Hathaway, Lady Gaga, Richard Gere, Paul McCartney, Julianne Moore, Graham Nash, Gwenth Paltrow, Bonnie Raitt, Tim Robbins, Todd Rundgren, Susan Sarandon, Ringo Starr, Martha Stewart, Uma Thurman, Liv Tyler, and Bethany Yarrow. Among the groups that are a part of the Artists group are the Beastie Boys, Black Keys, Indigo Girls, Patti Smith Group, The B-52s, and White Out.

Of Americans with wide-spread name recognition, Robert F. Kennedy Jr. is among the most vocal opponents of fracking. A month before New York's Department of Environmental Conservation appointed him to an advisory panel on fracking, Kennedy charged:

> "[T]he natural gas industry has been reckless and irresponsible and dishonest with the American public and they've lost much of their credibility. . . .
> "[B]ecause of the lack of candor by the industry, because of

their reckless behavior, it's unclear whether we can get that natural gas out of the ground without causing cataclysmic environmental damage."[758]

Protests and more than 80,000 public comments,[759] along with Kennedy's strong opposition, may have convinced Gov. Cuomo, who had been leaning to allow fracking under certain circumstances, to continue the four-year moratorium and conduct yet another study, specifically directed to investigating the public health effects.[760]

Energy in Depth (EID), the oil and gas industry's PR front, which uses personal attacks as a prime method of rebuttal, had previously attacked Kennedy when it unsuccessfully tried to ingratiate itself with the residents of the gas fields: "[They] have put up with a continuous barrage of insults and hyperbole over the last three years as . . . Bobbie Kennedy, Jr. and like-minded charlatans have abused them in the name of natural gas obstructionism."[761] Apparently, EID erroneously believed that using both a diminutive and feminine form of Kennedy's first name somehow defused his credibility. EID claimed Sean Lennon was nothing more than a "struggling artist" who was "using an anti-natural gas agenda to promote himself."[762] It attacked the group of high-profile musicians, actors, and others in the creative arts, most of whom had long histories of involvement in social issues, as "artists looking for relevance" who were merely "regurgitating outdated and tired information."[763]
However, it was outdated information that EID tried to slip past the unknowing public. In one of many accusations against the claims of anti-fracking activists, EID claimed their evidence of a high failure rate of cement well casings was faulty, and there were only 184 failures in more than 34,000 wells.[764] But, EID's data was for 1983 to 2007, before fracking began in the Marcellus Shale. Strangling logic, EID claimed that without fracking struggling dairy farmers would not be able to reap the financial rewards of having gas wells on their fields.
"Artists who have turned into activists on this issue are being led by those who are more comfortable twisting the facts and taking part in street theater, stunts and gimmicks," said Brad Gill, president of the industry-front group Independent Oil and Gas Association of New York (IOGA). Reciting the energy

industry's dogma of prosperity through fracking, Gill said the Artists "are ignoring the prosperity and environmental protection that modern natural gas development is bringing to many other states."[765]

On Jan. 17, 2013, Artists Against Fracking took a chartered bus into Pennsylvania's Susquehanna County to look at well sites and talk with the people who were affected by the drillers. Trailing the celebrities were the media and representatives of the natural gas industry. Rebecca Roter, of Kingsley, Pa., one of the activists, explained:

> "The celebrities had come . . . to hear the voices that have been marginalized by media, by industry spin, by the state of Pennsylvania. They did not come for a sound bite or to take a photograph of flaming tap water—they wanted to reach out to us, to the people living with fracking, and to see for themselves how our lives have been affected by shale gas extraction. . . .
>
> "Yoko Ono, Sean Lennon, Susan Sarandon, and Arun Gandhi [grandson of Mohandas Gandhi] accomplished more in one day to draw national attention to the human cost of shale gas extraction than our collective voices have in six years of confronting natural gas development in Susquehanna County. Local news reports about the celebrities' tour of Susquehanna County gasfields finally included the voices and personal stories of local residents speaking about the human cost of shale extraction as they have experienced it. The press had to cover that story because the celebrities about whom they thronged were listening to us"[766]

After talking with residents, Sarandon said, "If it's been decided that these people are expendable, and that the people in this area are expendable, there's nothing to stop (the industry) from thinking that they can sacrifice other people in other places."[767]

Arun Gandhi concluded, "We are committing violence against nature, against resources, against environment and eventually this is going to destroy us, destroy humanity."[768]

As expected, EID had its "take" on the bus tour. Tom Shepstone said the celebrities "just come up here to pick on this area and use it as part of their trendy cause."[769] He made sure the media also heard his opinion that "It's definitely a celebrity tour, it's a stunt, and I have to wonder if it's not connected with Sean Lennon releasing an album yesterday."[770]

Gasland, Promised Land, and Other Media

The presence of methane in drinking water in Dimock, Pa., had become the focal point for Josh Fox's investigative documentary, *Gasland,* which received the Sundance Special Jury Prize and then began airing on HBO in June 2010. Robert Koehler, writing in *Variety,* called it "one of the most effective and expressive environmental films of recent years. [It] may become to the dangers of natural gas drilling what *Silent Spring* was to DDT."[771] The 107–minute film received an Academy Award nomination in 2011 for Outstanding Documentary; Fox also received an Emmy for non-fiction directing.

Hundreds of activist organizations have shown the film, many with Fox as guest speaker, to rally the people already opposed to fracking and to inform those who had limited knowledge about the process about the effects upon their health and community.

Fox's interest in fracking had intensified when a natural gas company offered $100,000 for mineral rights on about 20 acres of property his family owned in Milanville, in the extreme northeast part of Pennsylvania, about 60 miles east of Dimock.

The natural gas industry launched a major public relations campaign opposing *Gasland,* and tried to deny it Academy Award honors. Rex W. Tillerson, CEO of ExxonMobil, declared, "The 'Gasland' movie did more to set us back in this endeavor [to discount fear about fracking] than anything else out there, and yet every aspect of that movie has been completely, scientifically debunked."[772] Energy in Depth created a seven page rebuttal.[773] However, Josh Fox effectively responded to EID's attacks in his own 39-page counter-rebuttal.[774]

Gasland is a film "Joseph Goebbels would have been proud," said Teddy Borawski, chief oil/gas geologist for the Pennsylvania Department of Conservation and Natural Resources, who called *Gasland* "propaganda" and compared Fox, whose father and paternal grandparents survived the Holocaust, to the Nazi propaganda minister.[775] Within a week, a contrite Borawski issued a public apology, acknowledging he "used very poor judgment," and that his remarks "do not truly reflect the person that I am or my understanding of the atrocities committed by the Nazis."[776]

John Hanger, Gov. Ed Rendell's DEP secretary who was interviewed by Fox, opposed *Gasland*. In his blog for Feb. 23, 2011, Hanger claimed the film "presents a selective, distorted view of gas drilling and the energy choices America faces today [and] seeks to inflame public opinion to shut down the natural gas industry."[777] Hanger, an advocate of natural gas drilling and fracking, later declared he was a candidate to replace Tom Corbett as governor.

Chesapeake Energy backed a 35-minute film rebuttal, *Truthland*, which claimed Fox was "a spoiled avant-garde showman from New York City."[778] That pretend-documentary focused upon Shelly Depue, a farmer from Susquehanna County, who went on what was billed as an independent national fact-finding tour; she concluded that fracking was safe. Although the film emphasized Depue was not paid, it skirted the issue that wells were put onto her farm, for which she and her family would receive several forms of income. As *Littlesis.org,* the website for the Public Accountability Initiative, noted in its own investigation, "[T]he film and its 'full-scale website and social media campaign' [pushed by EID] was planned from start to finish by the natural gas industry."[779] The original website[780] was registered in the name of Chesapeake Energy on Feb. 1, 2012, but later re-assigned to the Independent Petroleum Association of America. *Littlesis* points out that the film, with all the characteristics of an infomercial, "was made by a high-profile ad firm based in Hollywood, CA and promoted by a Washington DC media relations expert."[781] Several of EID's senior staff had previously been employed by Republican members of Congress, according to Dory Hippauf,[782] whose "Connects the Dots" series, published in *Common Sense 2,*[783] shows relationships between EID and other natural gas industry front groups, industry funding, and politics.

On his Facebook Page, July 17, 2012, State Rep. Jesse White (D-Washington County), wrote that following a screening:

> "Industry representatives told the 211 people in attendance the gas industry is going to start getting more aggressive in attacking people who don't agree with them, and that because their facts are totally right, there can be no middle ground in

the debate. So much for that 'honest and fact-based' discussion we always hear about but rarely see, huh? Maybe there was more truth revealed than I realized tonight."[784]

Audiences could also try to find the truth in Josh Fox's 18-minute video, *The Sky is Pink*, a plea to Gov. Andrew Cuomo to continue the moratorium on fracking in New York; or in the industry-friendly *FrackNation*, directed and produced by conservatives Phelim McAleer and Ann McElhinney. McAleer says the film tells the truth about the benefits of fracking; McElhinney said her team "talked to people across the country in areas where fracking is happening in their backyard. They all told us the same thing: *Fracking is safe and it's saving our community, but the media is ignoring us.* We were willing to listen."[785] The film premiered on the AXS cable network in January 2013. Among their previous films was *Not Evil, Just Wrong* (2009), which tried to debunk the Oscar-winning *An Inconvenient Truth* (2006), a documentary about Al Gore's campaign to bring environmental awareness to the people.

Gasland, a documentary, grossed $30,846[786] in its first (and only) month of theatrical release; the rest of the income is primarily from television and video sales and rentals. The first major motion picture to focus upon the natural gas industry brought in about $173,915 in box office receipts the first week of a limited release beginning Dec. 28, 2012.[787] By the end of its first weekend of general release (Jan. 4–6, 2013), *Promised Land* grossed $4.6 million from 1,676 theaters, before it began fading from general release after two weeks. By the end of the third week, it had grossed $7.6 million.[788]

With a $15 million budget,[789] *Promised Land* was written by Matt Damon and John Krasinski, based upon a story by Dave Eggers, and directed by Gus Van Sant. The movie stars Damon as a sweet-talking salesman from a natural gas driller who is trying to lease drilling rights from families in a town hit hard by the recession, Krasinski as an environmentalist, and Hal Holbrook as a teacher who is knowledgeable by a lifetime of experience. The movie was partly filmed in western Pennsylvania in Spring 2012.

"We wanted to show how it [fracking] tears apart local communities and subverts democracies and corrupts political leaders and eviscerates all the things that Americams value," Damon told Robert F. Kennedy Jr.[790]

Even before its release, *Promised Land* was attacked by a "rapid response team" from Energy in Depth, and by a horde of conservatives, including Fox News,[791] the *National Review,*[792] and the website, Breitbart.com.[793] Most of the objections focused upon claims that the movie, a drama not a documentary, is an amalgamation of factually-inaccurate liberal bias that claims fracking is unsafe. Cherry-picking wisps of information from a few research studies and an orchard of half-truths from the natural gas industry, the right-wing press claimed not only is there no danger in the water, but that any claims about environmental damage and health problems are exaggerations meant solely to fuel a liberal bias.

The movie "lacks any substantive scientific evidence illustrating the alleged dangers of fracking [and] falls back on many conventional anti-capitalist themes," wrote Bill Bennett,[794] conservative political commentator and former secretary of education in the Reagan administration. Bennett's solution to the energy problem is "freeing up more state and federal lands for drilling [and] cutting unnecessary, burdensome restrictions."

Energy in Depth created a Facebook website, "The Real Promised Land," to present what it called:

> "real life narratives of how farmers, homeowners and business owners, living on shale gas deposits have been able to realize the benefits of natural gas development in their towns and communities. Their genuine experiences with hydraulic fracturing reveal that the process isn't unregulated or dangerous, but rather a means for economic growth and job creation, in areas of the country that would have otherwise been economically barren."[795]

Nevertheless, the movie justifiably received mixed reviews from media critics, primarily because of its descent into clichés in characterization and weakness in the script. But, there was another problem that hovered over the film.

Three months before the film's release, the conservative Heritage Foundation, which had said the creators "have gone to absurd lengths to vilify oil and gas companies," now revealed that financing came from a wholly-owned subsidiary of the United Arab Emirates, "a member of the Organization of Petroleum Exporting Countries (OPEC), [which] has a stake in the future of the American fossil fuel industry." Thus, the Heritage Foundation concluded, the reason to fund *Promised Land* was "slowing the development of America's natural gas industry" to protect the petroleum industry.[796] The story generated hundreds of negative stories by conservative commentators and others who favored fracking. One of those was John Hanger.

Hanger's pro-industry review said the movie is "grating, arrogant, elitist [and] sends the horribly unfair, false message that those who sign drilling leases typically are greedy, stupid, and waste their gains on conspicuous consumption like sports cars." Several times in his review, Hanger referred to the film's "Persian Gulf investors." The development of gas production and distribution in the United States, said Hanger, "threatens the power of oil dictators in the Middle East and Putin's Russia that has a near monopoly on supplying gas to Europe. These oil and gas oligarchs, therefore, use their bulging purses to assault shale gas production that could mean new gas production in many countries and a lessening of their geopolitical power."[797]

What the Heritage Foundation, Hanger, and others didn't reveal was that neither the UAE nor its media subsidiary funded the film. In 2004, billionaire Jeff Skoll, the first president of eBay, formed Participant Media to produce socially relevant films. Among them were *An Incovenient Truth, American Gun, Fast Food Nation, The World According to Sesame Street, Syriana,* and *Good Night and Good Luck.* In 2008, Image Nation of Abu Dhabi invested $250 million into the company to assist in production of several films of social significance.[798] Image Nation Abu Dhabi did not have editorial control and, according to Participant Media, the investment "covers all qualifying Participant narrative films regardless of genre or subject matter."[799] Damon said the first time he and Krasinski "were aware that Image Nation was involved with our movie was when we saw the rough cut and saw their logo."[800]

However, there was another "bulging purse." The Marcellus Shale Commission produced and placed a 15-second ad before every showing of *Promised Land* in about three-fourths of all Pennsylvania theaters.[801] What the Commission may not have factored into what would normally be a brilliant marketing strategy was that most of the audience were probably already opposed to fracking, weren't susceptible to the politically biased negative reviews, and saw the movie to get reinforcement for their own beliefs. The fact the movie had artistic flaws did not negate the reality that anti-fracking arguments were receiving national attention and discussion.

Pamphlets and books, music and theatre, wrap themselves around all social movements. Among major books to explore the social issues of fracking are those from Bill McKibben, Vikram Rao, Seamus McGraw, Frank R. Spellman, and Tom Wilber. More than two dozen songs look at the effects of fracking. Among the singer/songwriters, most of whom have posted their music on YouTube, are Peter Alsop, Marc Black, Anne Hills, David Holmes and Niel Bekker, Bill Hunter, Joel Kalma, Kris Kitko, Corey Koehler, Judith Van Allen, Ben Warren.

Dr. Stephen Cleghorn, a sociologist and farmer from Jefferson County, Pa., uses a PowerPoint presentation to enhance his speeches about problems with fracking and its impact upon local communities.

"Same River," produced by New York City's Strike Anywhere Performance Ensemble, has been touring the Marcellus region since summer 2010. The multimedia production combines a structured script with improvisational acting, dance, music, and stage lighting and video which are displayed on several surfaces. The production "is a reflection of the places we perform," says Leese Walker, artistic director and Ensemble producer. Each performance is an extended three-act experience. The first act is a local work of art that hangs in the lobby; Act 2 is the 75-minute structured improv; Act 3 is a town hall discussion. "Our intent," says Walker, "is not only to address a socio-political issue but also to initiate dialogue, to spur the people to think, talk, and act."

The enthusiasm and determination of activists doing what they can to protect the environment and public health—while demanding that politicians, governments, and corporations think about the future and not act for immediate gratification—has only grown stronger as more people learn the facts and have become part of what will undoubtedly be known as one of the major social movements of the past decade.

CHAPTER 14
A Brief Look into the Future

During the first six months of 2012, the energy industry mined 895 billion cubic feet of gas in Pennsylvania by fracking the earth, slightly more than twice what it produced the previous year.[802] About 85 percent of production came from six counties (Bradford, 26 percent; Susquehanna, 21 percent; Lycoming, 11 percent; Tioga, 10 percent; Washington, 9 percent; and Greene, 8 percent.)[803] Five companies—Anadarko, Chesapeake, Cabot, Range Resources, and Talisman—accounted for about 60 percent of all gas mined by unconventional wells. The overproduction of gas, combined with a mild winter in 2011–2012, led to a gas glut, with the lowest prices in 10 years—and happy homeowners.

Not as happy were the natural gas companies that saw profits plunge, with several companies taking losses because of over-drilling and their greedy rush to production. "Because of the intricate financial deals and leasing arrangements that many of them struck during the boom, they were unable to pull their foot off the accelerator fast enough to avoid a crash in the price of natural gas,[804] which is down more than 60 percent since the summer of 2008," wrote *New York Times* reporters Clifford Krauss and Eric Lipton in October 2012.[805] The energy exploration and distribution companies, they noted, "are committed to spending far more to produce gas than they can earn selling it. Their stock prices and debt ratings have been hammered."[806]

Rex Tillerson, CEO of ExxonMobil, was blunt in his assessment: "We are all losing our shirts today. We're making no money. It's all in the red."[807]

The oil and gas industry is facing an even greater burden from a coal industry that has no plans to fade away. Coal generates about half of all electricity in the U.S.[808] Between 2000 and 2010, coal as a source of electricity rose more than 56 percent throughout the world, with a prediction of an increase of 40 percent between 2010 and 2020.[809] IEA projects that oil will be second most used energy source, at 23 percent; and natural gas will be third, at 23 percent by 2035.[810] Projections by the International Energy Agency (IEA) are that by 2035, coal, with 30 percent of the world market, will replace oil as the leader among sources of energy generation.

Helping to keep coal as a major energy source is the development of clean coal technology, which reduces, but doesn't eliminate, the emission of nitrogen oxide, carbon and sulfur dioxides, and other particulates and gases that had once made coal the dirtiest of all energy sources and a major contributor to ozone depletion and the destruction of the environment. The use of coal as a source of energy is almost as great as the use of gas, nuclear, water, oil, solar, and wind power combined. However, oil, coal, and gas resources are finite, and alternative non-fossil fuel sources will increase during the next decade.

Nevertheless, no matter how clean coal, oil, or gas energy become, the processes to mine and refine them are dirty, and workers are exploited. The United Mine Workers represents about 42 percent of all employed coal miners;[811] the gas drilling industry, for the most part, is non-union. The workers, therefore, have no bargaining or grievance rights; health and workplace benefits for workers who aren't executives or professsionals, are minimal.

The Sierra Club argues that the country needs "to leapfrog over gas whenever possible in favor of truly clean energy [wind, solar, water]. Instead of rushing to see how quickly we can extract natural gas, we should be focusing on how to be sure we are using less—and safeguarding our health and environment in the meantime."[812] Millions throughout the world are in agreement. Wind turbines provide about one-fifth of all power in Iowa and South Dakota.[813] Two-thirds of the world's increase in solar energy in 2011 was in Europe;[814] a two mile section of track between Paris and Amsterdam allows trains to run entirely on solar power;[815] in May 2012, Germany's solar energy plants produced as much power as 20 nuclear power plants.[816] In the United States, solar panel costs have plunged more than 80 percent in five years. Energy writer Maria Gallucci predicts by the end of 2013, the U.S. will have 7,600 megawatts of solar capacity, the equivalent of almost two and a half large nuclear plants.[817] The U.S., says Gallucci, has also developed about "50,000 megawatts worth of wind turbines [which] power 13 million American homes."[818]

A significant increase of non-fossil fuel energy would produce a cleaner fuel source, reduce workplace accidents and substantially negate the health and environmental effects of fossil fuel energy; it would also increase employment, one of the basic reasons why politicians say they like natural gas drilling.

To compensate for their own overproduction, the mild winter, and competition from the coal industry, natural gas energy companies developed a three-part strategy: reduce production, look for better revenue-producing areas, and export gas.

To correct for overproduction, natural gas companies reduced production to drive prices back up. Aubrey McClendon, CEO of Chesapeake Energy, which took 21 percent of the mined gas in the first half of 2012, said "We are now projecting a decline in . . . gas production of approximately 7% in 2013."[819]

Matt Kelso of fracktracker.org said the data confirms, "The industry is ramping down permitting, drilling and production efforts across the State for the time being."[820] Many companies are "holding gas back in anticipation of a better price, and if that happens, they can turn the wells back on very quickly," said Dr. Terry Engelder, professor of geosciences at Penn State.[821] Dr. Sergei Komlev, head of export contracts and pricing for the Russian-owned Gazprom, the world's largest energy company, told the Associated Press, "We do not expect the currently abnormally low prices in the USA to last for long."[822]

The severe cut-back in production may benefit the energy companies, but it has caused a problem within local industry; workers are laid off and local businesses, many that raised prices to take advantage of the better-paid work force and enjoyed a brief surge in income, are now struggling in many areas.

To accomplish the second part of their strategy, energy companies are looking to move back into oil production, where prices continue to stay high, or expand their operations into the deeper Utica Shale in Ohio, mining the more lucrative wet gas than the dry gas in the Marcellus Shale. Mike Knapp, president of Knapp Acquisitions & Production, explains the differences:

211

"Natural gas is a gas comprised of multiple hydrocarbons, the most prevalent being methane. The higher the methane concentration, the 'drier' or 'colder' the gas is. Other constituents of natural gas are evaporated liquids like ethane and butane, pentane, etc. We refer to these collectively as natural gas liquids (NGLs), or 'condensates'. The higher the percenttage of NGL's, the 'hotter' or 'wetter' the gas is. NGL's must be stripped out of the gas before it can be put in a pipeline and used. Ethane, which is prevalent in Western PA wet gas, is the feedstock for Ethylene, which is what we use to make plastics."[823]

As wet gas begins to dominate production, economic benefits of dry gas will diminish:

"Whether you are in a wet gas or dry gas area is going to have a huge impact on the value of your lease. Right now, the NGL's are worth considerably more than dry gas. In some areas, the value of the gas is more than doubled because of the NGL's. . . . Dry gas areas are dead as a doornail right now for leasing, but wet gas areas are seeing nice offers. With the impressive . . . results companies have been having in the Utica in Ohio with oil production (which is far more profitable than wet or dry gas) dry gas areas have been reduced to a distant third tier. Dry gas areas will not be in high demand for a long time, possibly decades. That is not to say that they will not be drilled . . . but companies will not be competing and landowners shouldn't expect to see the huge up front bonuses (that they did a few years ago) again any time soon. With the low price of gas, it's simply not economical to pay out thousands of dollars per acre just to be able to pull a rig on the property to spend millions to drill a well that will barely make a profit at these prices. Wet gas area landowners have ... more leverage."[824]

The third strategy is exporting. The Fukushima Daiichi meltdown in Japan has made countries re-evaluate their dependence upon nuclear energy. "Given the uncertain prospects of nuclear power in Japan, the political drive to clean up the Chinese energy system and the acute energy shortages in India, Asia is intensely looking for energy supplies," Maria van der Hoeven, executive director of the International Energy Agency, told a business audience in Calgary in August 2012.[825]

The U.S. Energy Information Administration estimates that in the next decade production of natural gas will exceed consumption, rising from about 5.0 trillion cubic feet in 2010 to about 8 trillion cubic feet in 2020.[826] This, according to EIA estimates, will lead to the U.S. becoming "a net exporter of liquefied natural gas (LNG) in 2016" to take advantage of lower gas prices in the U.S. while being able to command higher prices overseas. Projections for 2035, according to the EIA, are for 13.6 trillion cubic feet of production.

To help assure both a market and higher prices overseas, the energy exploration and distribution corporations have a plan. The American Petroleum Institute, the Independent Petroleum Institute, the U.S. Chamber of Commerce, and the National Foreign Trade Council sued the Securities and Exchange Commission in October 2012 to reduce oversight of corporate finances and practices.[827] The plaintiffs want the federal courts to declare unconstitutional, under the First Amendment, a regulation within the Dodd–Frank Act of 2010 that requires energy companies to disclose on their annual reports any payments made to foreign governments and their leaders. The purpose of that regulation, as well as the Foreign Corrupt Practices Act of 1977, which the plaintiffs also protested, is to eliminate bribes paid to get and keep contracts. "Transparency is . . . taking a beating as U.S. firms fight for the right to bribe foreign governments and hide their activities from American shareholders and the citizens of the nations where they do business," observed Sara Jerving and Mary Bottari, writing in *PR Watch*, a web-based newsletter published by the Center for Media and Democracy.[828]

More than 100 of the nation's leading physicians, scientists, and environmental engineers petitioned President Obama in December 2012 to reconsider increased exportating of natural gas. The petition pointed out that the "opening of LNG export facilities would serve to accelerate fracking in the United States in absence of sound scientific assessment, placing policy before health."[829]

Dr. Seth B. Shonkoff, an environmental scientist and executive director of the group that had petitioned the President, asked:

"Why would the United States dramatically increase the use of an energy extraction method without first ensuring that the trade-off is not the health of Americans in exchange for the energy demands of foreign nations? The only prudent thing to do here is to conduct the needed research first."[830]

Even if the U.S. rethinks approvals for overseas distribution, natural gas producers may find that estimates of overseas purchases may not match optimistic predictions. China, India, and most European countries have been developing non-fossil green energy sources faster than the United States.

While significant reduction in drilling to bring back higher prices for natural gas and increased exporting may give stockholders a better return on investment (ROI), an underlying truth remains—the government and industry's claims of economic benefits of natural gas for 40 years are suspect. "With these new shale gas wells, you get peak production in one or two years, when the pressure is strongest, then a steep fall-off," says geologist Sally Odland of Columbia University's Lamont–Doherty Earth Observatory.[831] Estimates of 30–40 years of well production "remain unproven [and] has significantly stressed the financials of the industry as payback based on long productive output is not yet a proven part of shale gas dynamics," says Robert Magyar, one of the nation's leading experts in energy production.[832] The only reason the gas companies are drilling in parts of the Marcellus Shale, says Odland, is "to hold position and keep their leases."[833]

Financial analyst Deborah Rogers, whose specialty is issues involving the natural gas industry, told the *Village Voice* in September 2012:

"After a decade of fracking, we're beginning to be able to show that, without a doubt, this was simply a very well-orchestrated public relations campaign. . . . There is gas there, but is there as much as they said? No. Are we gonna see the economic stability they promised? The answer is no." . . .

"Shale gas was supposed to be this economic powerhouse for the next 40 years, they said. It didn't even work out in the past seven. And it's the same story in every other state. Unfortunately, that's just how the game is played."[834]

Like the sub-prime mortgage crisis that began about 2006

214

and led to the largest recession since the Great Depression in the early 1930s, hydraulic horizontal fracturing in the Marcellus Shale, with its myriad government concessions, may eventually prove to be more beneficial to the industry than to the consumers.

It's not unreasonable to expect people who are unemployed or underemployed to grasp for anything to help themselves and their families, nor is it unreasonable to expect that persons— roustabouts, clerks, truck drivers, safety inspectors, helicopter pilots, and engineers among several hundred thousand in dozens of job classifications—will take better paid jobs, even if it often means 60–70 hour work weeks under hazardous conditions. It's also not unreasonable to expect that families living in agricultural and rural areas, who are struggling to survive, will snap at the lure of several thousand dollars to lease mineral rights and some of their land to an energy company, which will also pay royalties. But what is unreasonable is that government willingly issues corporate welfare assistance in the form of tax inducements and other benefits to allow corporations to flourish at the expense of the people and their environment.

Conclusion

When the history of natural gas exploration in Pennsylvania is finally written, the story will be that for a few years it was a cheaper, cleaner-burning energy source than coal or oil, that it temporarily helped some people in rural areas, and brought some well-paying jobs into the state. But history will also write that as with every other energy industry that moved into the region, when the natural gas industry finally abandons the Marcellus Shale, having taken all it could, it will leave thousands of dry wells dotting the landscape, with many of the wells possibly leaking methane and wastewater, and that the lure of immediate gratification led Pennsylvania's politicians to willingly accept political donations that led them to sacrifice their citizens' health and the state's environment.

PHOTO: Frank Finan

Workers, without HazMat protection, clean up fluids near Lathrop Twp., Pa.

ENDNOTES

[1] http://stateimpact.npr.org/pennsylvania/2012/06/29/dep-secretary-krancer-stumps-for-natural-gas-processing-facility-in-delaware-county/
[2] http://www.weitzlux.com/Marcus-Hook-Pa_1961415.html
[3] http://epa-sites.findthedata.org/l/495723/Sunoco-Inc-r-And-m-Marcus-Hook-Refinery
[4] http://stateimpact.npr.org/pennsylvania/2012/07/11/despite-refinery-closure-petrochemical-plant-remains-in-delaware-county/
[5] http://www.aga.org/Kc/analyses-and-statistics/statistics/annualstats/appliance/Documents/Table10-4.pdf
[6] http://www.ngvjournal.dreamhosters.com/en/statistics/item/911-worldwide-ngv-statistics
[7] http://www.cngnow.com/vehicles/pages/information.aspx
[8] http://www.afdc.energy.gov/fuels/stations_counts.html
[9] http://stateimpact.npr.org/pennsylvania/2012/01/24/energy-in-tonights-state-of-the-union/
[10] http://belfercenter.ksg.harvard.edu/files/The Geopolitics of Natural Gas.pdf
[11] http://www.bakerinstitute.org/publications/EF-pub-HKSGeopoliticsOfNaturalGas-073012.pdf
[12] http://www.csmonitor.com/Environment/Latest-News-Wires/2012/1001/Natural-gas-boom-in-US.-Is-Russia-the-big-loser
[13] http://thetimes-tribune.com/news/business/u-s-gas-boom-rattles-russia-1.1381170
[14] http://www.magazine.columbia.edu/print/1091
[15] http://www.newsolutionsjournal.com/
[16] http://www.cnn.com/2012/08/29/us/new-york-fracking-artists-protest/index.html
[17] http://water.epa.gov/type/groundwater/uic/class2/hydraulicfracturing/wells_hydrowhat.cfm
[18] http://www.portal.gov.on.ca/drinkingwater/stel01_049392.pdf
[19] http://www.njgeology.org/enviroed/newsletter/v2n1.pdf
[20] http://nyagainstfracking.org/no-compromise-on-an-independent-comprehensive-health-impact-assessment/
[21] http://smartenergyuniverse.com/utility-news/923-epa-issues-final-air-rules-for-the-oil-and-natural-gas-industry
[22] http://www.huffingtonpost.com/robert-f-kennedy-jr/fracking-natural-gas-new-york-times-_b_1022337.html
[23] http://www.dec.ny.gov/energy/75370.html.
[24] http://ecowatch.org/2012/water-for-fracking/
[25] http://money.cnn.com/2012/08/10/news/economy/kansas-oil-boom-drought/index.html
[26] http://www.nytimes.com/2012/09/06/us/struggle-for-water-in-colorado-with-rise-in-fracking.html?pagewanted=all
[27] http://www.nytimes.com/2012/09/06/us/struggle-for-water-in-colorado-with-rise-in-fracking.html?pagewanted=all
[28] http://www.r-cause.net/uploads/8/0/2/5/8025484/frackingwithpropane1.pdf

[29] http://www.epa.gov/ghgreporting/documents/pdf/2010/Subpart-W_TSD.pdf
[30] http://www.envirobank.org/index.php?sid=5&rec=138
[31] http://www.endocrinedisruption.com/files/Oct2011HERA10-48forweb3-3-11.pdf
[32] http://energy.nationaljournal.com/2010/09/natural-gas-a-fracking-mess.php
[33] http://www.ewg.org/analysis/usgs-recent-earthquakes-almost-certainly-manmade
[34] http://www.worldwatch.org/despite-methane-emissions-upstream-natural-gas-cleaner-coal-life-cycle-basis
[35]

https://www.paoilandgasreporting.state.pa.us/publicreports/Modules/DataExports/DataExports.aspx
[36]

https://www.paoilandgasreporting.state.pa.us/publicreports/Modules/DataExports/DataExports.aspx
[37] Theuticashale.com/fracking-our-food-supply
[38] http://www.ewg.org/analysis/usgs-recent-earthquakes-almost-certainly-manmade
[39] http://www.eia.gov/totalenergy/data/annual/pdf/sec6_11.pdf
[40] http://www.nytimes.com/2011/02/27/us/27gas.html?_r=2&pagewanted=all
[41] http://www.scientificamerican.com/article.cfm?id=safety-first-fracking-second
[42] http://www.eia.gov/forecasts/aeo/pdf/0383er(2011).pdf
[43] http://www.usnews.com/science/news/articles/2012/09/23/decades-of-federal-dollars-helped-fuel-gas-boom
[44] http://thebreakthrough.org/archive/interview_with_dan_steward_for
[45] http://bigstory.ap.org/article/decades-federal-dollars-helped-fuel-gas-boom
[46] http://www.eia.gov/analysis/studies/worldshalegas/
[47] http://www.eia.gov/forecasts/aeo/er/
[48] http://geology.com/articles/marcellus-shale.shtml
[49] http://nyshalegasnow.blogspot.com/2010/11/contentious-bedrock-photos-of-marcellus.html
[50] http://www.nps.gov/frhi/parkmgmt/upload/GRD-M-Shale_12-11-2008_high_res.pdf
[51] http://www.growthstockwire.com/3148/Shocking-Gov-t-Statistics-Every-Investor-Should-See
[52] http://www.eia.gov/forecasts/aeo/er/
[53] http://newsok.com/marcellus-natural-gas-production-expanded-in-2012/article/feed/479753
[54] http://www.pressconnects.com/article/20110925/NEWS01/109250340/Gas-face-off-set-DEC-public-hearing
[55] http://gasfreeseneca.com/?page_id=123
[56] http://www.dcbureau.org/201206157397/natural-resources-news-service/epa-refuses-to-release-seismic-data.html
[57] http://ir.eia.gov/ngs/ngs.html
[58]

http://www.eia.gov/pub/oil_gas/natural_gas/analysis_publications/ngpipeline/undrgrnd_storage.html

59 http://theadvocate.com/home/4227993-125/scientists-give-sinkhole-insight
60 http://www.nola.com/environment/index.ssf/2012/08/sinkhole_neighbors_dont_know_w.html
61 http://theadvocate.com/news/4282059-123/more-land-and-trees-fall
62 http://goingglobaleastmeetswest.blogspot.com/2012/12/louisiana-sinkhole-evacuees-wont-be.html
63 http://www.sunherald.com/2012/11/17/4310905/vent-wells-burning-gas-from-louisiana.html
64 http://www.examiner.com/article/sinkhole-strong-chemical-stench-on-bubbling-bayou
65 http://www.eia.gov/dnav/ng/ng_stor_sum_dcu_spa_m.htm
66 http://ec.europa.eu/environment/integration/energy/unconventional_en.htm
67 http://www2.canada.com/calgaryherald/iphone/business/calgary/story.html?id=6107167
68 http://www.upi.com/Science_News/2011/04/21/South-Africa-halts-gas-fracking-plan/UPI-23521303431387/
69 http://gdacc.org/2012/06/15/daily-maverick-confessions-of-a-fracking-defector/
70 http://af.reuters.com/article/investingNews/idAFJOE88601L20120907
71 http://www.dailykos.com/story/2010/11/30/924263/-Fracking-A-New-York
72 http://reuters_th.adam.com/content.aspx?productId=16&pid=16&gid=52790
73 http://www.huffingtonpost.com/2012/07/02/north-carolina-governor-fracking-perdue_n_1642004.html
74 http://www.reuters.com/article/2009/12/23/energy-fracking-newyork-idUSN2220711920091223?type=marketsNews
75 http://www.usatoday.com/money/industries/energy/2010-11-16-nat-gas-ban-pittsburgh_N.htm
76 http://2politicaljunkies.blogspot.com/2011/06/what-frack-catholic-cemeteries-gas.html
77 http://blogs.philadelphiaweekly.com/phillynow/2011/01/25/city-council-to-vote-on-drilling-moratorium-green-party-response/
78 http://www.psehealthyenergy.org/data/LNG_SignOnLetterPDF.pdf
79 http://www.marcellusgas.org/graphs/PA#pcnty
80 http://www.dced.state.pa.us/public/oor/constitution.pdf
81 http://www.marcellusgas.org/graphs/PA#prodollars
82 http://files.dep.state.pa.us/OilGas/BOGM/BOGMPortalFiles/OilGasReports/2012/WEBSITE Weekly Report for Last Week.pdf
83 http://stateimpact.npr.org/pennsylvania/2011/09/12/can-pennsylvanias-state-forests-survive-additional-marcellus-shale-drilling/
84 http://explorepahistory.com/story.php?storyId=1-9-20&chapter=1
85 http://www.eia.gov/beta/state/?sid=PA
86 http://www.nytimes.com/2011/03/09/opinion/lweb09gas.html

87
http://www.eia.gov/electricity/monthly/epm_table_grapher.cfm?t=epmt_1_12_a
88 http://heartland.org/issues/environment
89 http://www.usnews.com/science/news/articles/2012/09/23/decades-of-federal-dollars-helped-fuel-gas-boom
90 http://www.epa.gov/lawsregs/laws/rcra.html
91 http://www.epa.gov/superfund/policy/cercla.htm
92 http://www.nytimes.com/2011/03/04/us/04gas.html?pagewanted=all
93 http://www.rff.org/rff/documents/rff-dp-01-38.pdf
94 *Ibid.*
95
http://permanent.access.gpo.gov/lps21800/www.epa.gov/safewater/uic/cbmstudy.html
96
http://www.earthworksaction.org/files/publications/Weston.pdf?pubs/Weston.pdf
97 http://www.gpo.gov/fdsys/pkg/PLAW-109publ58/pdf/PLAW-109publ58.pdf
98 http://water.epa.gov/grants_funding/dwsrf/index.cfm
99
http://www.google.com/url?sa=t&rct=j&q=&esrc=s&frm=1&source=web&cd=1&ved=0CCUQFjAA&url=http%3A%2F%2Fwww.wildwatch.org%2FBinocular%2Fbino25%2FHydro-fracturingImpactonWildlif.doc&ei=neRlT4T-DYmJgwfws7XKAg&usg=AFQjCNHhsrEhZunrz78hXtCTrLMJ0PFXog&sig2=0imb2JYsl
100 http://grist.org/?p=143090&preview=true
101 http://www.huffingtonpost.com/robert-f-kennedy-jr/fracking-natural-gas-new-york-times-_b_1022337.html
102
http://www.earthworksaction.org/files/publications/PetroleumExemptions1c.pdf?pubs/PetroleumExemptions1c.pdf
103 http://www.epa.gov/region1/nepa/
104 http://www.epa.gov/superfund/policy/cercla.htm
105 http://www.scientificamerican.com/article.cfm?id=safety-first-fracking-second
106 *Ibid.*
107 http://thomas.loc.gov/cgi-bin/bdquery/z?d111:H.R.2766:
108 http://thomas.loc.gov/cgi-bin/bdquery/z?d111:S1215:
109 http://www.govtrack.us/congress/bills/112/hr1084
110 http://www.govtrack.us/congress/bills/112/s587
111 http://www.businessweek.com/news/2012-03-29/obama-says-oil-company-profits-justify-ending-u-dot-s-dot-tax-br
112 http://www.businessweek.com/news/2012-03-29/obama-says-oil-company-profits-justify-ending-u-dot-s-dot-tax-br
113 *Ibid.*
114 http://thinkprogress.org/green/2012/03/29/454853/senators-who-voted-to-protect-oil-tax-breaks-received-23582500-from-big-oil/

[115] http://www.whitehouse.gov/the-press-office/2012/04/13/statements-president-s-executive-order-supporting-safe-and-responsible-d

[116] http://www.whitehouse.gov/the-press-office/2012/04/13/statements-president-s-executive-order-supporting-safe-and-responsible-d

[117] http://thehill.com/blogs/e2-wire/e2-wire/221429-industry-groups-applaud-obamas-natural-gas-executive-order

[118] http://www.whitehouse.gov/state-of-the-union-2013

[119] http://www.opensecrets.org/news/2011/05/big-companies-special-interests-hire-private-congressional-delegations-to-lobby.html

[120] http://www.opensecrets.org/industries/indus.php?Ind=E01

[121] *Ibid.*

[122]
http://www.opensecrets.org/industries/recips.php?ind=E01&cycle=2012&recip detail=P&mem=N&sortorder=U

[123] http://www.american.com/archive/2012/august/presidential-power-obama-vs-romney-on-energy

[124] *Ibid.*

[125] Ibid.

[126]
http://www.opensecrets.org/industries/recips.php?ind=E01&cycle=2008&recip detail=P&mem=N&sortorder=U

http://www.opensecrets.org/industries/recips.php?ind=E01&cycle=200 8&recipdetail=P&mem=N&sortorder=U

[127] *Ibid.*

[128]
http://www.opensecrets.org/industries/summary.php?ind=E01&recipdetail=M &sortorder=U&cycle=2012

[129]
http://www.opensecrets.org/industries/summary.php?ind=E01&cycle=2012&r ecipdetail=H&sortorder=N&mem=Y

[130]
https://www.opensecrets.org/politicians/industries.php?cycle=2012&cid=N000 25495&type=I&newmem=N

[131] http://dailyitem.com/0100_news/x62498341/Barletta-on-shale-caucus-owns-gas-stock

[132] *Ibid.*

[133] *Ibid.*

[134]
http://www.opensecrets.org/politicians/industries.php?type=C&cid=N0000558 2&newMem=N&cycle=2012

[135]
http://epw.senate.gov/public/index.cfm?FuseAction=PressRoom.PressReleases &ContentRecord_id=7B837DA0-802A-23AD-43CD-4C7F2458F87B

[136]
http://epw.senate.gov/public/index.cfm?FuseAction=Minority.PressReleases& ContentRecord_id=7280f114-802a-23ad-4bb2-797081f70515&Region_id=&Issue_id=

137 http://thehill.com/blogs/e2-wire/e2-wire/205455-white-house-throws-in-the-towel-on-interior-nominee

138 http://www.canadafreepress.com/index.php/article/44055

139 http://www.eenews.net/public/Greenwire/2012/01/20/1

140 http://sierraclub.typepad.com/compass/2012/02/the-sierra-club-and-natural-gas.html

141 *Ibid.*

142 http://www.sierraclub.org/sierra/201207/pennsylvania-fracking-shale-gas-200-sidebar.aspx

143 http://supreme.justia.com/cases/federal/us/558/08-205/

144 http://www.whitehouse.gov/the-press-office/statement-president-todays-supreme-court-decision-0

145 http://legaltimes.typepad.com/blt/2010/01/obama-supreme-court-opened-the-floodgates-for-special-interests.html

146 http://query.nictusa.com/cgi-bin/dcdev/forms/C00504530/828309/sa/ALL

147 http://blog.seattlepi.com/seattlepolitics/2012/10/28/2-5-million-from-chevron-usa-to-republicans/

148 http://marcellusmoney.org/sites/default/files/images/Marcellus Money press release 071212_0.pdf

149 http://marcellusmoney.org/sites/default/files/images/Marcellus Money press release 071212_0.pdf

150 http://stateimpact.npr.org/pennsylvania/2011/11/10/common-cause-report-details-campaign-contributions-from-drillers/

151 http://marcellusmoney.org/sites/default/files/images/Marcellus Money press release 071212_0.pdf

152 http://www.marcellusprotest.org/sites/marcellusprotest.org/files/December_20 12.pdf

153 http://articles.philly.com/2011-06-29/news/29717481_1_corbett-campaign-tom-corbett-marcellus-shale

154 http://www.rollingstone.com/politics/news/the-big-fracking-bubble-the-scam-behind-the-gas-boom-20120301

155 http://www.marcellusgas.org/graphs/PA#prodollars

156 http://www.marcellusgas.org/record_book_co.php?report_type=top_producing _co&county_id=&num_results=10&muni_id=

157 http://www.triplepundit.com/2011/03/socially-responsible-investment-firm-divests-chesapeake-energy/

158 http://www.marcellusgas.org/record_book_co.php?report_type=top_inspection s_co&county_id=&num_results=10&muni_id=

159 http://thepennsylvaniaprogressive.com/diary/3232/tom-corbett-same-old-corruption

160 http://www.propublica.org/article/corbett-pa-energy-exec-authority-environment

161 http://www.portal.state.pa.us/portal/server.pt?open=512&objID=708&PageID =224602&mode=2&contentid=http://pubcontent.state.pa.us/publishedcontent/

publish/cop_general_government_operations/oa/oa_portal/omd/p_and_p/execu
tive_orders/2010_2019/items/2010_05.html
[162] http://www.marcellus.psu.edu/resources/PDFs/MSACFinalReport.pdf.
[163] https://stateimpact.npr.org/pennsylvania/2012/02/07/governors-budget-a-
mixed-bag-for-conservationists/
[164] http://change.nature.org/2011/02/10/how-pennsylvania%E2%80%99s-
energy-infrastructure-will-affect-hunters-fishers-trout-birds/
[165]
http://blog.pennlive.com/midstate_impact/print.html?entry=/2012/10/state_pa
rks_director_john_norb.html
[166] http://stateimpact.npr.org/pennsylvania/tag/tom-corbett/
[167] http://articles.philly.com/2012-09-20/business/33978477_1_corbett-
pennsylvania-s-marcellus-shale-shale-gas-outrage
[168] http://www.propublica.org/article/corbett-pa-energy-exec-authority-
environment
[169] http://www.theintelligencer.net/page/content.detail/id/567362/Pa--Still-
Seeking--Cracker-.html?nav=515
[170] http://www.gastruth.org/?p=163
[171] http://marcellusprotest.org/dep-inspectors-limited-propublica
[172] *Ibid.*
[173] http://thetimes-tribune.com/opinion/dep-boss-bows-to-gas-drillers-
1.1126421
[174] http://old.post-gazette.com/pg/11123/1143606-503-0.stm
[175]
http://www.legis.state.pa.us/CFDOCS/Legis/PN/Public/btCheck.cfm?txtType=
HTM&sessYr=2011&sessInd=0&billBody=H&billTyp=B&billNbr=1950&pn=
3048
[176]
http://www.legis.state.pa.us/CFDOCS/Legis/RC/Public/rc_view_action2.cfm?s
ess_yr=2011&sess_ind=0&rc_body=H&rc_nbr=854
[177] http://legiscan.com/gaits/rollcall/103497
[178] http://www.facebook.com/photo.php?v=10200406919639519
[179] http://www.newpa.com/newsroom/governor-corbett-signs-historic-
marcellus-shale-law
[180] http://www.gastruth.org/?p=534
[181] http://marcellusmoney.org/sites/default/files/images/Marcellus Money
press release 071212_0.pdf
[182] http://wmarcellusmoney.org/candidates
[183] http://marcellusmoney.org/sites/default/files/images/Marcellus Money
press release 071212_0.pdf
[184] http://old.post-gazette.com/pg/12115/1226722-178-0.stm
[185] http://articles.philly.com/2012-02-06/news/31030487_1_guilty-verdicts-
corbett-bonusgate
[186] http://www.heraldstandard.com/gcm/news/top_stories/deweese-others-
speak-out-against-corbett-plan/article_8f577f62-d7ae-5131-b859-
92ff99e0e6f5.html
[187] http://www.bloomberg.com/news/2012-07-23/frackers-fund-university-
research-that-proves-their-case.html

[188] http://marcelluscoalition.org/about/
[189] http://www.bloomberg.com/news/2012-07-23/frackers-fund-university-research-that-proves-their-case.html
[190] http://www.businessweek.com/ap/financialnews/D9MA9IF80.htm
[191] http://pennbpc.org/sites/pennbpc.org/files/2009-natural-gas-production-ranking-and-2010-11-drilling-tax-rates.pdf
[192] http://thirdandstate.org/2012/february/pa-marcellus-shale-fee-among-lowest-nation
[193] http://youngphillypolitics.com/topics/natural_gas_drilling
[194] http://marcellusmoney.org/sites/default/files/images/Marcellus Money press release 071212_0.pdf
[195] http://standardspeaker.com/news/downtown-rally-blasts-spending-plan-1.1126526
[196] http://www.philly.com/philly/news/politics/state/20120910_ap_padrillingimpactfeeraisesmorethan200m.html
[197] http://www.statejournal.com/story/18154011/pa-gas-drilling-brought-35-billion-in-2011
[198] http://stateimpact.npr.org/pennsylvania/2012/02/08/corbetts-budget-would-cut-dep-spending/
[199] http://www.pennlive.com/editorials/index.ssf/2012/04/gov_corbetts_budget_hurts_envi.html
[200] Ibid.
[201] http://paindependent.com/2012/03/dep-budget-cut-not-affecting-pa-gas-well-checks/
[202] http://www.nytimes.com/2012/05/15/us/for-oil-workers-deadliest-danger-is-driving.html?_r=3
[203] http://www.portal.state.pa.us/portal/server.pt/gateway/PTARGS_0_2_785_708_0_43/http%3B/pubcontent.state.pa.us/publishedcontent/publish/global/files/executive_orders/2010___2019/2012_11.pdf
[204] http://www.businessweek.com/ap/2012-08-13/foes-pa-dot-state-permit-order-threatens-environment
[205] http://www.usatoday.com/money/industries/energy/2011-04-13-pa-gas-drilling-permits.htm
[206] http://www.earthworksaction.org/files/publications/FINAL-US-enforcement-sm.pdf
[207] Ibid.
[208] http://earthjustice.org/news/press/2012/groups-urge-penn-governor-to-reverse-policy-that-delays-warning-of-fracking-water-pollution
[209] http://www.legis.state.pa.us/cfdocs/billinfo/billinfo.cfm?syear=2011&sind=0&body=H&type=B&BN=1659
[210] http://fracktoids.blogspot.com/2012/06/hb-1659-ez-frack-permits.html
[211] http://dpwsd.waterworld.com/WaterWorld/en-us/index/display/elp-article-tool-template._printArticle.waterworld.world-

regions.europe.2011.12.Wastewater-plants-not-designed-for-fracking-water-says-Robert-F-Kennedy-Jr.html

[212] http://www.texassharon.com/2011/11/28/who-put-the-k-in-fracking-the-truth-the-whole-truth-and-nothing-but-the-fracking-truth/

[213] http://www.mainlinemedianews.com/articles/2011/04/29/main_line_suburban_life/news/doc4db9eae2a7f7c181410251.txt

[214] http://www.marcellusmoney.org/candidates

[215] http://www.legis.state.pa.us/cfdocs/billinfo/billinfo.cfm?syear=2011&sind=0&body=S&type=B&bn=367

[216] http://www.post-gazette.com/stories/news/environment/drilling-on-campus-marcellus-shale-boom-puts-colleges-at-crossroads-322630/?print=1

[217] *Ibid.*

[218] http://mansfield.edu/marcellus-institute/events/marcellus-summer-camp/

[219] http://www.ccp.edu/site/news_room/press_releases/2012/111412.php

[220] http://protectingourwaters.wordpress.com/2012/11/28/breaking-faculty-group-stands-up-to-marcellus-shale-coalition-at-community-college-of-philadelphia/#comments

[221] http://protectingourwaters.wordpress.com/2012/12/06/aft-union-calls-fracking-risks-dangers-unacceptable-urges-clean-energy-jobs-training/

[222] http://www.post-gazette.com/stories/news/environment/drilling-on-campus-marcellus-shale-boom-puts-colleges-at-crossroads-322630/?print=1

[223] http://www.shell.us/aboutshell/projects-locations/appalachia.html

[224] http://publicsource.org/investigations/potter-township-forgotten-player-bringing-shell-oil-pa

[225] http://triblive.com/news/2008446-74/jobs-state-plant-permanent-shell-industry-corbett-numbers-tax-billion#axzz2DbxQ6ptB

[226] http://www.shell.us/home/content/usa/aboutshell/media_center/news_and_press_releases/2012/03152012_pennsylvania.html

[227] http://www.facebook.com/photo.php?v=10200406919639519

[228] http://www.legis.state.pa.us/cfdocs/billinfo/billinfo.cfm?syear=2011&sind=0&body=S&type=B&bn=1263

[229] http://pubs.usgs.gov/fs/2012/3075/fs2012-3075.pdf

[230] http://www.phillyburbs.com/my_town/palisades/pa-lawmakers-approve-gas-drilling-moratorium-for-bucks-and-montco/article_5ece9717-8d8e-5e87-ae87-2737c134a187.html

[231] http://www.phillyburbs.com/my_town/palisades/pa-lawmakers-approve-gas-drilling-moratorium-for-bucks-and-montco/article_5ece9717-8d8e-5e87-ae87-2737c134a187.html

[232] http://www.post-gazette.com/stories/local/state/oil-and-gas-drilling-permits-on-hold-for-southeastern-pa-642767/

[233] *Ibid.*

[234] http://www.gastruth.org/?p=534

[235] http://www.phillyburbs.com/my_town/palisades/pa-lawmakers-approve-gas-drilling-moratorium-for-bucks-and-montco/article_5ece9717-8d8e-5e87-ae87-2737c134a187.html

[236] http://www.phillyburbs.com/my_town/palisades/pa-lawmakers-approve-gas-drilling-moratorium-for-bucks-and-montco/article_5ece9717-8d8e-5e87-ae87-2737c134a187.html

[237]

http://www.buckslocalnews.com/articles/2012/07/01/bristol_pilot/news/doc4ff0
67a5cea73798138351.txt

[238] http://www.commoncause.org/site/pp.asp?c=dkLNK1MQIwG&b=8482079

[239] *Ibid.*

[240] http://www.tomlibous.com/index.asp?type=B_BASIC&SEC={8273960E-5FCD-4429-BAE6-98818C88E676}

[241] http://www.commoncause.org/site/pp.asp?c=dkLNK1MQIwG&b=8482079

[242] *Ibid.*

[243]

http://www.opensecrets.org/politicians/summary.php?cid=N00030949&cycle=
2012

[244]

http://www.legis.state.pa.us/CFDOCS/Legis/PN/Public/btCheck.cfm?txtType=
HTM&sessYr=2011&sessInd=0&billBody=H&billTyp=B&billNbr=1950&pn=
3048

[245] http://money.cnn.com/2012/05/01/news/economy/fracking-violations/index.htm

[246] *Ibid.*

[247] *Ibid.*

[248] http://www.examiner.com/article/gov-corbett-says-yes-to-shale-gas-incentives-but-no-to-solar

[249] http://www.americanlegislator.org/2012/03/alec-encourages-responsible-resource-production/

[250] http://www.scientificamerican.com/article.cfm?id=drill-for-natural-gas-pollute-water&print=true

[251] http://www.counterpunch.org/2012/03/19/the-perils-of-fracking/

[252] http://www.nejm.org/doi/full/10.1056/NEJMsb1209858

[253] http://hosted2.ap.org/PAWIC/APUSnews/Article_2012-04-19-Gas Drilling-Health/id-d48d0a70cde6443c9f105279abfac99f

[254] http://www.samsmithpahouse.com/NewsItem.aspx?NewsID=14132

[255] http://nurses.3cdn.net/39c3056f1d418b5a7f_xfm6bkbib.pdf

[256] http://www.courthousenews.com/2012/07/31/48847.htm

[257] *Ibid.*

[258] http://www.denverpost.com/opinion/ci_18436002

[259] http://truth-out.org/news/item/7323:fracking-pennsylvania-gags-physicians

[260] http://www.cedclaw.org/wp-content/uploads/2012/04/Middlefield-Complaint1.pdf

[261] http://drydensec.org/sites/default/files/AnschutzMemorandumOfLaw.pdf

[262] http://www.dailycamera.com/erie-news/ci_20126684/noaa-study-erie-gas-drilling-moratorium-fracking-propane-butane

263 http://www.dailycamera.com/erie-news/ci_20126684/noaa-study-erie-gas-drilling-moratorium-fracking-propane-butane
264 http://ecowatch.org/2012/fighting-fracking-gold-rush/
265 http://www.post-gazette.com/stories/local/marcellusshale/pittsburgh-inspired-colo-towns-fracking-ban-665637/
266 http://www.nytimes.com/2012/11/26/us/with-ban-on-fracking-colorado-town-lands-in-thick-of-dispute.html?_r=0
267

http://ballotpedia.org/wiki/index.php/Longmont_City_Fracking_Ban_Questio
n_(November_2012)
268 http://www.bizjournals.com/denver/news/2012/12/06/hickenlooper-colorado-wont-sue.html
269 *Ibid.*
270 http://caselaw.findlaw.com/pa-supreme-court/1144542.html
271 http://www.puc.state.pa.us/naturalgas/naturalgas_marcellus_Shale.aspx
272 http://www.delawareriverkeeper.org/resources/Comments/FINAL
PETITION PART 1.pdf
273

http://www.alternet.org/story/154459/fracking_democracy:_why_pennsylvania
%27s_act_13_may_be_the_nation%27s_worst_corporate_giveaway_?akid=839
1.298606.GhNYU8&rd=1&t=1
274

http://www.marcellusprotest.org/sites/marcellusprotest.org/files/June_2012.p
df
275 http://www.spilmanlaw.com/Mobile/Home/Resources/Attorney-Authored-Articles/Marcellus-Fairway/Pa--Commonwealth-Court-Strikes-Down-Act-13-Zoning
276 http://www.facebook.com/BerksGasTruth/posts/480779425267417
277 http://www.post-gazette.com/stories/local/marcellusshale/pennsylvania-submits-its-arguments-in-shale-drilling-law-appeal-651861/
278 http://www.post-gazette.com/stories/local/neighborhoods-south/local-group-challenges-new-rules-for-shale-gas-industry-651150/
279 http://canon-mcmillan.patch.com/articles/commonwealth-court-to-puc-cease-and-desist-release-the-impact-fee-money
280 http://stateimpact.npr.org/pennsylvania/2012/10/26/court-bars-public-utility-commission-from-reviewing-drilling-ordinances/
281 http://www.mcall.com/news/breaking/mc-gas-drilling-value-20120505,0,5829301.story
282

http://www.salon.com/2012/05/05/ap_pa_gas_drilling_brought_3_5_billion_in_
2011/
283 http://www.villagevoice.com/2012-09-19/news/boom-or-doom-fracking-environment/4/
284 http://www.businessweek.com/ap/2012-07-19/us-chamber-touts-benefits-of-pa-dot-gas-drilling
285 http://pipeline.post-gazette.com/news/archives/24704-chamber-officials-lobby-to-keep-shale-costs-low
286 https://www.opensecrets.org/lobby/top.php?showYear=a&indexType=s

287 https://www.opensecrets.org/lobby/top.php?showYear=2012&indexType=s
288 https://www.pachamber.org/newsroom/articles/2012/PA Chamber
president helps launch national Marcellus Shale campaign.php
289 http://www.energyindepth.org/tag/marcellus-shale-coalition/
290 http://marcelluscoalition.org/2012/09/msc-president-in-philadelphia-
inquirer-marcellus-shale-transforming-pa/
291 http://www.post-gazette.com/stories/local/marcellusshale/chamber-
officials-lobby-to-keep-shale-costs-low-645818/
292 http://www.earthworksaction.org/2010summit/Panel7_DeborahRogers.pdf
293 Ibid.
294 http://notohydrofracking.blogspot.com/2011/10/economic-assessment-of-
hydrofracking-dr.html
295 http://keystoneresearch.org/publications/research/drilling-deeper-job-
claims-actual-contribution-marcellus-shale-pennsylvania-jo
296 Ibid.
297 http://www.nytimes.com/2012/10/21/business/energy-environment/in-a-
natural-gas-glut-big-winners-and-losers.html?pagewanted=all&_r=3&
298 http://www.morningstar.com/earnings/PrintTranscript.aspx?id=13725415
299
http://aese.psu.edu/research/centers/cecd/publications/marcellus/marcellus-
shale-land-ownership-local-voice-and-the-distribution-of-lease-and-royalty-
dollars/view
300 http://www.desmogblog.com/2012/09/21/oil-and-gas-leases-create-conflicts-
fema
301 http://www.nationwide.com/newsroom/071312-FrackingStatement.jsp
302 http://www.fema.gov/pdf/nfip/manual201205/splashscreen.pdf
303 http://www.desmogblog.com/2012/09/21/oil-and-gas-leases-create-conflicts-
fema
304 http://www.desmogblog.com/2012/09/21/oil-and-gas-leases-create-conflicts-
fema
305 http://www.desmogblog.com/2012/09/21/oil-and-gas-leases-create-conflicts-
fema
306 http://www.nytimes.com/2012/09/30/realestate/fracking-fears-hurt-second-
home-sales-in-catskills.html?pagewanted=all&_r=1&
307 http://blog.360mtg.com/?p=2407
308 http://www.magazine.columbia.edu/features/summer-2012/gas-menagerie
309 http://www.rcalaw.com/Publications/Litigation/Condemnation-Issues-
Under-the-Natural-Gas-Act.php
310 http://uk.reuters.com/article/2009/05/03/btscenes-us-energy-gas-drilling-
idUKTRE5422TG20090503
311 http://www.tctimes.com/news/homeowner-s-property-destroyed-by-
enbridge/article_6ca1168e-e89f-11e1-97a0-001a4bcf887a.html
312 http://www.fwweekly.com/2009/10/14/sacrificed-to-shale/
313 http://www.latimes.com/news/nation/nationnow/la-na-nn-keystone-xl-
pipeline-texas-20120823,0,7657215.story
314 http://www.reuters.com/article/2012/10/02/us-chesapeake-landgrab-
substory-idUSBRE8910E920121002
315 http://www.supreme.courts.state.tx.us/historical/2012/mar/090901rh.pdf

[316] http://www.coloradoan.com/apps/pbcs.dll/article?AID=2012305180028
[317] http://www.westernresourceadvocates.org/schooldrill/
[318] *Ibid.*
[319] http://www.coloradoan.com/apps/pbcs.dll/article?AID=2012305180028
[320]

http://www.legis.state.pa.us/cfdocs/billinfo/billinfo.cfm?syear=2009&sind=0&
body=H&type=B&bn=977
[321]

http://www.pressconnects.com/viewart/20100811/NEWS11/8110334/Rendell-
willing-negotiate-gas-pooling-law
[322] http://www.bizjournals.com/pittsburgh/stories/2010/02/08/story12.html
[323]

http://www.legis.state.pa.us/cfdocs/billinfo/billinfo.cfm?syear=2011&sind=0&
body=S&type=B&bn=0447
[324] http://thetimes-tribune.com/news/forced-pooling-legislation-for-gas-
industry-planned-in-pennsylvania-1.885341
[325] http://www.propublica.org/article/forced-pooling-when-landowners-cant-
say-no-to-drilling
[326]

http://www.blm.gov/wo/st/en/prog/energy/oil_and_gas/best_management_prac
tices/split_estate.html
[327] http://www.bullfrogfilms.com/catalog/split.html
[328] *Ibid.*
[329]

http://www.businesswire.com/news/home/20120430006113/en/CORRECTING
-REPLACING-2000-Truck-Trips-Removed-PA
[330] http://aspe.hhs.gov/poverty/11poverty.shtml
[331]

https://www.aquaamerica.com/News/Pages/AquaAmericaReportsIncreaseinF
ourthQuarterandYear-EndEarnings.aspx
[332] http://www.saveriverdale.com/
[333] http://www.sungazette.com/page/content.detail/id/577215/Unfair.html
[334] http://www.sungazette.com/page/content.comment/id/575944/32-unit-
village-no-more.html?nav=5019
[335] http://articles.philly.com/2012-04-18/business/31361933_1_fracking-
trailer-park-anti-drilling/2
[336] http://articles.philly.com/2012-04-18/business/31361933_1_fracking-
trailer-park-anti-drilling
[337] http://www.ragingchickenpress.org/2012/06/07/hands-across-riverdale-
barricades-the-forward-facing-body-of-the-occupation/
[338] http://huffmaster.com/
[339] http://www.ragingchickenpress.org/2012/06/14/2569/
[340] http://www.saveriverdale.com/2012/06/riverdale-after-eviction_13.html
[341] http://www.sungazette.com/page/content.detail/id/579556/-Amicable-
resolution---reportedly-is-reached-with-last-home-residents.html
[342] http://www.gpo.gov/fdsys/pkg/USCODE-2009-title15/html/USCODE-2009-
title15-chap53-subchapI-sec2605.htm

343
http://www.legis.state.pa.us/cfdocs/billinfo/bill_history.cfm?syear=2011&sind=0&body=H&type=B&bn=1767

344
http://www.pmha.org/Portals/0/pdf/Act261ManufacturedHomeCommunityRightsActasamendedbyAct80.pdf

345
http://www.lehighvalleylive.com/bethlehem/index.ssf/2012/05/proposal_would_buy_time_for_mo.html

346
http://www.legis.state.pa.us/CFDOCS/Legis/RC/Public/rc_view_action2.cfm?sess_yr=2011&sess_ind=0&rc_body=H&rc_nbr=1397

347 http://www.garfield-county.com/public-health/documents/1   Complete HIA without Appendix D.pdf

348
http://democrats.energycommerce.house.gov/sites/default/files/documents/Hydraulic Fracturing Report 4.18.11.pdf

349 http://www.cdc.gov/chronicdisease/resources/publications/AAG/dcpc.htm

350 http://www.endocrinedisruption.com/chemicals.introduction.php

351
http://cce.cornell.edu/EnergyClimateChange/NaturalGasDev/Documents/PDFs/fracking chemicals from a public health perspective.pdf

352 http://gdacc.org/tag/drinking-water/

353 *Ibid.*

354 http://www.youtube.com/watch?v=N0on1TiO4DU&feature=youtu.be

355 http://www.epa.gov/iaq/schools/pdfs/publications/iaqtfs_update44.pdf

356 http://shale.sites.post-gazette.com/index.php/news/archives/24607-federal-agencies-probe-range-resources-yeager-marcellus-shale-gas-drilling-site

357 *Ibid.*

358
http://www.legis.state.pa.us/cfdocs/billinfo/billinfo.cfm?syear=2011&sind=0&body=S&type=B&BN=1100

359 http://www.syracuse.com/newsflash/index.ssf/story/residents-pa-ignoring-their-health-complaints/98ba5b39b40243659d47d0c42584b007

360
http://www.ombwatch.org/files/info/naturalgasfrackingdisclosure_highres.pdf

361 http://www.psehealthyenergy.org/data/lettertoGovCuomofinal.pdf

362 http://www.scribd.com/doc/67626513/Letter-to-Cuomo-on-Fracking

363 http://www.shalegas.energy.gov/resources/111811_final_report.pdf

364 http://psna.org/2012/06/nurses-promote-healthier-energy-choices/

365 http://earthjustice.org/sites/default/files/Hallowich_Brief.pdf

366 http://earthjustice.org/sites/default/files/Hallowich_Brief.pdf

367 http://earthjustice.org/

368 http://stateimpact.npr.org/pennsylvania/2012/12/07/appeals-court-agrees-with-newspapers-in-sealed-fracking-case/

369 http://stateimpact.npr.org/pennsylvania/2012/06/21/chesapeake-to-pay-1-6-million-for-contaminating-water-wells-in-bradford-county/

370

http://www.philly.com/philly/news/politics/state/20120622_ap_16msettlement inpagasdrillinglawsuit.html

[371] http://www.epa.gov/oecaagct/lcra.html

[372] http://www.ombwatch.org/node/12130

373

http://switchboard.nrdc.org/blogs/amall/new_nrdc_analysis_state_fracki.html

[374] *Ibid.*

[375] http://www.whitehouse.gov/state-of-the-union-2012

376

http://www.doi.gov/news/pressreleases/loader.cfm?csModule=security/getfile&pageid=293916

[377] http://www.chron.com/business/article/Q-amp-A-Openness-on-fracturing-ured-3797370.php

[378] http://fracfocus.org/

[379] http://stateimpact.npr.org/texas/2012/12/14/the-number-of-fracking-trade-secrets-in-texas-will-likely-surprise-you/

[380] http://www.bloomberg.com/news/2012-08-14/fracking-hazards-obscured-in-failure-to-disclose-wells.html

[381] http://www.eia.gov/neic/experts/natgastop10.htm

[382] http://ecowatch.org/2012/water-for-fracking/

[383] http://www.bloomberg.com/news/2012-08-14/fracking-hazards-obscured-in-failure-to-disclose-wells.html

[384] http://www.bloomberg.com/news/2012-08-14/fracking-hazards-obscured-in-failure-to-disclose-wells.html

[385] *Ibid.*

[386] *Ibid.*

[387] http://www.fracfocus.org/

[388] http://www2.canada.com/story.html?id=6763527

[389] http://dailyitem.com/0100_news/x1284938395/Susquehanna-River-Basin-Commission-approves-water-use-for-drilling

[390] http://stateimpact.npr.org/pennsylvania/2011/11/18/growing-tensions-within-the-delaware-river-basin-commission-halt-decision-on-gas-drilling/

391

http://www.portal.state.pa.us/portal/server.pt?open=514&objID=1072223&parentname=ObjMgr&parentid=396&mode=2

[392] http://articles.philly.com/2012-08-06/business/33049766_1_pennenvironment-michael-krancer-shale-gas

[393] http://articles.philly.com/2012-08-06/business/33049766_1_pennenvironment-michael-krancer-shale-gas

[394] http://news.nationalgeographic.com/news/2010/10/101022-energy-marcellus-shale-gas-environment/

[395] http://pafaces.wordpress.com/2010/04/23/stephanie-hallowich-speaks-out/

[396] http://www.marcellus-shale.us/Stephanie-Hallowich.htm

[397] http://www.marcellus-shale.us/Stephanie-Hallowich.htm

[398] http://pafaces.wordpress.com/2010/04/23/stephanie-hallowich-speaks-out/

[399] http://pafaces.wordpress.com/copyright-and-disclosure-policies/

[400] http://news.nationalgeographic.com/news/2010/10/101022-energy-marcellus-shale-gas-environment/
[401] *Ibid.*
[402] http://shale.sites.post-gazette.com/index.php/news/archives/24176-hallowich-family-files-court-action-against-range-resources
[403] http://protectingourwaters.wordpress.com/2012/09/17/methane-from-gas-drilling-manning-family-told-dont-use-your-kitchen-stove/
[404] http://www.ens-newswire.com/ens/feb2012/2012-02-16-02.html
[405] http://www.propublica.org/article/scientific-study-links-flammable-drinking-water-to-fracking
[406] http://www.scribd.com/doc/91428341/Delivered-Testimony-of-Professor-Margaret-Rafferty-R-N-at-Hydrofracking-Forum-4-25-12
[407] http://www.ipcc.ch/pdf/assessment-report/ar4/wg1/ar4-wg1-chapter2.pdf
[408] http://news.nationalgeographic.com/news/2010/10/101022-energy-marcellus-shale-gas-environment/
[409] http://www.casey.senate.gov/issues/issue/?id=ce690501-bb77-4195-a6d1-87152b9ae298
[410] http://www.dnr.state.oh.us/Portals/11/bainbridge/report.pdf
[411] *Ibid.*
[412] *Ibid.*
[413] http://www.oilandgaslawyerblog.com/Range Production Company Closing Statement.pdf
[414] http://www.dallasnews.com/health/medicine/20101209-epa-2-parker-county-homes-at-risk-of-explosion-after-gas-from-fracked_well-contaminates-aquifer.ece
[415] http://www.texaspolicy.com/pdf/Joint Stipulation of Dismissal.pdf
[416] http://citizenspeak.org/campaign/saynotofracking/epa-send-clean-water-families-impacted-fracking-butler-county-pa
[417] http://protectingourwaters.wordpress.com/2012/03/02/i-just-want-water-demonstrators-confront-rex-energy-in-butler-county/
[418] http://www.marcellusoutreachbutler.org/2/post/2012/03/the-plethora-of-excuses-and-explanations-disintegrates.html
[419] http://citizensvoice.com/news/drilling/dep-fines-chesapeake-1-1-million-for-fire-contamination-incidents-1.1148249
[420] http://wcexaminer.com/index.php/archives/news/30232
[421]
http://www.cleanair.org/program/outdoor_air_pollution/marcellus_shale/independent_study_finds_significant_fault_line_methane
[422] https://www.dropbox.com/s/kircmcdy7jdtfgw/DEP Letter to Minott 071212.pdf
[423]
http://www.boston.com/news/local/massachusetts/articles/2012/09/07/report_finds_methane_remains_issue_in_pa_township/
[424] *Ibid.*
[425] http://thetimes-tribune.com/news/dep-secretary-methane-may-have-leaked-through-perforations-in-bradford-gas-well-1.1343005

426
http://www.boston.com/news/local/massachusetts/articles/2012/09/07/report_fi nds_methane_remains_issue_in_pa_township/?page=2
427 http://www.timesleader.com/stories/Dimock-Twp-property-owners-sue-gas-driller-Cabot,106231
428 http://www.propublica.org/article/officials-in-three-states-pin-water-woes-on-gas-drilling-426
429 http://weeklypress.com/shale-shame-cabot-fined-heavily-for-dimock-water-contamination-p1896-1.htm
430 http://thetimes-tribune.com/news/dimock-officials-reject-offer-of-water-deliveries-1.1241292
431 http://dailyitem.com/0100_news/x431310713/Cabot-CEO-EPA-investigation-of-Dimock-water-wastes-taxpayer-money
432 http://thetimes-tribune.com/news/dep-head-calls-epa-knowledge-of-dimock-rudimentary-1.1255658
433 http://stateimpact.npr.org/pennsylvania/2012/06/01/krancer-once-again-tells-washington-to-back-off/
434 http://www.ohio.com/news/break-news/pennsylvania-governor-says-drilling-opponents-are-unreasoning-1.336049
435 http://ecowatch.org/2012/epa-finds-water-safe-to-drink-despite-explose-levels-of-methane-and-other-toxins/
436 http://www.propublica.org/article/so-is-dimocks-water-really-safe-to-drink
437 http://waterdefense.org/blog/water-defense-cries-foul-epa-statement
438 http://waterdefense.org/news/so-dimock%E2%80%99s-water-really-safe-drink
439 http://citizensvoice.com/news/cabot-and-dimock-families-near-settlement-1.1358912
440 http://citizensvoice.com/news/cabot-and-dimock-families-near-settlement-1.1358912
441 http://wnep.com/2012/08/22/dep-cabot-allowed-to-frack-in-dimock/
442 http://thetimes-tribune.com/news/dep-lets-cabot-resume-dimock-fracking-1.1361871
443
http://www.marcellusgas.org/record_book_co.php?report_type=top_inspection s_co&county_id=&num_results=10&muni_id=
444 http://www.epa.gov/aboutepa/states/dimock-atsdr.pdf
445 *Ibid.*
446
http://www.atsdr.cdc.gov/hac/pha/ChesapeakeATGASWellSite/ChesapeakeAT GASWellSiteHC110411Final.pdf
447 http://thetimes-tribune.com/news/after-blowout-most-evacuated-families-return-to-their-homes-in-bradford-county-1.1135253
448 http://thetimes-tribune.com/news/after-blowout-most-evacuated-families-return-to-their-homes-in-bradford-county-1.1135253
449 http://www.propublica.org/article/response-to-pa-gas-well-accident-took-13-hours-despite-state-plan-for-quick
450 *Ibid.*

[451] http://m.thedailyreview.com/news/chesapeake-gets-dep-notice-of-violation-1.1136716

[452] http://m.citizensvoice.com/news/drilling/gas-well-blowout-spills-frack-fluids-onto-farm-streams-in-bradford-county-1.1135482

[453] http://www.oag.state.md.us/Press/2011/050211.html

[454] http://www.bloomberg.com/article/2012-06-14/a0r9O1j1J9Rk.html

[455] http://theintelligencer.net/page/content.detail/id/575435/Chesapeake-to-Pay--600K-Fine-for-Filling-Wetzel-Co--Stream.html?nav=515

[456] http://dl.dropbox.com/u/48182083/drilling/nov.pdf

[457] http://www.post-gazette.com/stories/local/state/state-charges-local-company-for-dumping-wastewater-and-sludge-287538/?print=1

[458] http://www.alternet.org/fracking/toxic-wastewater-dumped-streets-and-rivers-night-gas-profiteers-getting-away-shocking?page=0%2C12&paging=off

[459] http://old.post-gazette.com/pg/pdf/201103/20110317_shipman_awws_gjpresentment.pdf

[460] http://old.post-gazette.com/pg/pdf/201103/20110317_shipman_awws_gjpresentment.pdf

[461] http://www.post-gazette.com/stories/local/state/attorney-general-critical-of-pollution-sentence-wastewater-668583/

[462] http://www.alternet.org/fracking/toxic-wastewater-dumped-streets-and-rivers-night-gas-profiteers-getting-away-shocking?page=0%2C12&paging=off

[463] http://www.pennfuture.org/UserFiles/Daze/20120416_Letter_DEPFBC_SusqSmallmouthBass.pdf

[464] Ibid.

[465] Ibid.

[466] http://www.paenvironmentdigest.com/newsletter/default.asp?NewsletterArticleID=22483

[467] https://docs.google.com/file/d/0B4Y3VQLxjkxOWjM0QTlua2tnYkE/edit?pli=1

[468] http://www.propublica.org/article/injection-wells-the-poison-beneath-us

[469] http://blogs.artvoice.com/avdaily/2012/07/20/ub-the-buffalo-news-bamboozled-by-natural-gas-industry/

[470] Ibid.

[471] http://www.srsi.buffalo.edu/wp-content/uploads/2012/05/UBSRSI-Environmental-Impacts-Single-Page.pdf

[472] http://www.desmogblog.com/2012/09/19/frackademia-the-brewing-suny-buffalo-shale-resources-society-institute-storm

[473] http://public-accountability.org/2012/05/ub-shale-play/

[474] http://www.buffalonews.com/apps/pbcs.dll/article?AID=/20121119/CITYANDREGION/121119113/1010

[475] http://www.rural.palegislature.us/documents/reports/Marcellus_and_drinking_water_2012.pdf

[476] Ibid.

[477] http://energy.utexas.edu/images/ei_shale_gas_regulation120215.pdf

[478] *Ibid.*

[479] http://public-accountability.org/wp-content/uploads/ContaminatedInquiry.pdf

[480] http://www.bloomberg.com/news/2012-07-23/frackers-fund-university-research-that-proves-their-case.html

[481] http://stateimpact.npr.org/texas/2012/07/24/texas-professor-on-the-defensive-over-fracking-money/

[482] *Ibid.*

[483] http://www.bizjournals.com/houston/morning_call/2012/12/university-of-texas-making-changes-in.html

[484] http://www.chron.com/news/houston-texas/houston/article/Panel-criticizes-fracking-study-UT-ethics-rules-4098257.php

[485] http://www.bloomberg.com/news/print/2012-12-06/texas-energy-institute-head-quits-amid-fracking-study-conflicts.html

[486] http://www.statesman.com/news/news/opinion/the-case-against-fracking/nRNQj/

[487] http://pennbpc.org/sites/pennbpc.org/files/CMSC-Final-Report.pdf

[488] http://www.energyindepth.org/wp-content/uploads/2012/05/myers-potential-pathways-from-hydraulic-fracturing4.pdf

[489]
http://s3.amazonaws.com/propublica/assets/methane/garfield_county_final2.pdf

[490] http://www.propublica.org/article/officials-in-three-states-pin-water-woes-on-gas-drilling-426

[491] http://www.sustainableotsego.org/Risk Assessment Natural Gas Extraction-1.htm

[492] *Ibid.*

[493] http://www.nicholas.duke.edu/cgc/pnas2011.pdf

[494] http://www.pnas.org/content/early/2012/07/03/1121181109.full.pdf+html

[495] http://today.duke.edu/2012/07/marcellus

[496]
http://www.epa.gov/region8/superfund/wy/pavillion/EPA_ReportOnPavillion_Dec-8-2011.pdf

[497] http://www.scientificamerican.com/article.cfm?id=fracking-linked-water-contamination-federal-agency

[498] http://docs.nrdc.org/energy/files/ene_12050101a.pdf

[499] *Ibid.*

[500] http://www.bloomberg.com/news/2012-09-26/diesel-compounds-found-in-water-near-wyoming-fracking-site-2-.html

[501] http://www.eveningtribune.com/opinions/columnists/x912153418/New-fracking-study-should-be-required-reading-for-local-leaders

[502] http://articles.baltimoresun.com/2012-01-04/features/bal-cdc-scientist-urges-more-gas-drilling-study-20120104_1_shale-gas-drilling-fracking-impacts

[503] http://www.epa.gov/airquality/oilandgas/basic.html

[504] *Ibid.*

[505] http://www.stateoftheair.org/2012/health-risks/health-risks-ozone.html

[506] http://www.eurekalert.org/pub_releases/2012-03/uocd-ssa031612.php

[507] http://www.sciencedirect.com/science/article/pii/S0048969712001933

[508]
http://www.springer.com/earth+sciences+and+geography/meteorology+%26+c
limatology?SGWID=0-10009-12-565099-0

[509] http://www.nirs.org/radiation/radonmarcellus.pdf

[510] http://www.nature.com/news/air-sampling-reveals-high-emissions-from-gas-field-1.9982

[511] http://researchmatters.noaa.gov/news/Pages/COoilgas.aspx

[512] http://www.wtop.com/?nid=41&sid=2651787

[513] *Ibid.*

[514] http://www.propublica.org/article/science-lags-as-health-problems-emerge-near-gas-fields

[515] http://protectingourwaters.wordpress.com/2012/08/16/shale-gas-industry-harms-air-quality-public-health-shale-gas-outrage-news-bulletin-2/

[516]
http://www.fwweekly.com/index.php?option=com_content&view=article&id=5
433:vapors-sicken-arlington&catid=76:metropolis&Itemid=377

[517] *Ibid.*

[518] *Ibid.*

[519] *Ibid.*

[520] http://www.fwweekly.com/2009/10/14/sacrificed-to-shale/

[521] http://townofdish.com/objects/DISH_-_final_report_revised.pdf

[522] *Ibid.*

[523] *Ibid.*

[524]
http://www.alternet.org/water/150794/trailer_talk's_frack_talk%3A_why_a_m
ayor_was_forced_to_leave_his_town_because_of_gas_drilling_/?page=entire

[525] http://www.post-gazette.com/stories/local/marcellusshale/hunting-club-contends-with-spring-water-contaminations-from-gas-drilling-300746/?print=1

[526] http://stateimpact.npr.org/texas/2012/01/20/railroad-commission-responds-to-explosion-in-pearsall/

[527]
http://www.pittsburghlive.com/x/pittsburghtrib/news/regional/s_724557.html

[528] http://shale.sites.post-gazette.com/index.php/news/archives/24437-compressor-station-explosion-shuts-down-at-least-10-wells

[529] http://www.papipelinesafety.org/news/archive/2012/01/

[530] http://citizensvoice.com/news/drilling/susquehanna-county-compression-station-up-and-running-without-state-permission-1.1295296

[531] http://www.naturalgaswatch.org/?p=1305

[532] http://www.philly.com/philly/news/special_packages/inquirer/marcellus-shale/135273768.html

[533] http://www.puc.state.pa.us/naturalgas/Act_127_Info.aspx

[534] http://articles.philly.com/2012-04-08/news/31308559_1_gas-safety-gas-explosion-natural-gas

[535] http://jurist.org/dateline/2012/07/garrett-eisenhour-pipeline-regulation.php

[536]http://www.postgazette.com/stories/local/marcellusshale/wva-blast-heightens-concerns-over-natural-gas-pipelines-666131/

[537] http://www.propublica.org/article/pipelines-explained-how-safe-are-americas-2.5-million-miles-of-pipelines

[538] http://www.csmonitor.com/Environment/2012/1212/West-Virginia-gas-pipeline-explosion-just-a-drop-in-the-disaster-bucket

[539] http://www.mercurynews.com/ci_16051111?source=most_emailed&nclick_check=1

[540]http://www.huffingtonpost.com/2010/09/12/san-bruno-explosion-photos_n_713976.html#138569

[541] http://www.mercurynews.com/ci_16051111?source=most_emailed&nclick_check=1

[542]http://www.mercurynews.com/ci_16045798?IADID=Search-www.mercurynews.com-www.mercurynews.com

[543] http://www.sfgate.com/bayarea/article/PG-E-diverted-safety-money-for-profit-bonuses-2500175.php

[544] http://www.democracynow.org/2012/8/8/chevron_oil_refinery_fire_in_richmond

[545] http://www.democracynow.org/2012/8/8/chevron_oil_refinery_fire_in_richmond

[546] http://www.mercurynews.com/breaking-news/ci_21302478/investigators-set-comb-through-chevron-richmond-fire-site

[547] http://www.democracynow.org/2012/8/8/chevron_oil_refinery_fire_in_richmond

[548] http://articles.latimes.com/2012/aug/14/opinion/la-oe-0814-juhasz-chevron-refinery-pollution-20120814

[549] http://www.huffingtonpost.com/2012/12/11/sissonville-west-virginia-explosion_n_2279577.html?ir=Green

[550] http://www.wvpubcast.org/newsarticle.aspx?id=27911

[551] http://newsok.com/wv-gas-explosion-comes-amid-line-shut-off-debate/article/feed/476071

[552] http://www.gao.gov/assets/590/589514.pdf

[553] *Ibid.*

[554] http://www.workers.org/2010/us/fracking_1021

[555] http://www.trb.org/HMCRP/HMCRPOverview.aspx

[556] http://www.alternet.org/fracking/toxic-wastewater-dumped-streets-and-rivers-night-gas-profiteers-getting-away-shocking?paging=off

[557] http://www.dispatch.com/content/stories/local/2012/07/12/speed-not-cause-of-derailment-officials-say.html

[558] http://wnep.com/2012/07/06/acid-spill-at-gas-well/

[559] http://wnep.com/2012/09/26/crash-spills-fluid-from-gas-drilling-into-creek/

[560] http://dailyitem.com/0100_news/x685074962/People-flee-as-chemical-cloud-hovers-near-New-Columbia

[561] *Ibid.*
[562] http://protectingourwaters.wordpress.com/2012/08/01/shale-gas-industry-puts-workers-at-risk-in-rush-to-frack/
[563] http://www.nytimes.com/2012/05/15/us/for-oil-workers-deadliest-danger-is-driving.html?_r=2&pagewanted=2
[564] http://protectingourwaters.wordpress.com/2012/07/18/whistle-blowing-truck-driver-on-law-flouting-fracking-companies/
[565] http://www.krextv.com/news/around-the-region/NC5-INVESTIGATION-Deadly-Gas-Cover-Up-Revealed-126869973.html
[566] *Ibid.*
[567] http://www.stopthefrackattack.org/wp-content/uploads/2012/07/STFA-BREATHE_Fact-Sheet1.pdf
[568] http://polis.house.gov/news/documentsingle.aspx?DocumentID=229905
[569] http://www.marcellus-shale.us/Marcellus_FRAC.htm
[570] http://ecowatch.org/2012/mining-companies-invade-wisconsin-for-frac-sand/
[571]
http://chippewa.com/search/?l=25&sd=desc&s=start_time&f=html&q=silica%20sand%20moratorium
[572] http://www.aflcio.org/content/download/.../1/.../safetyhealth_05222012.pdf
[573] http://blogs.cdc.gov/niosh-science-blog/2012/05/silica-fracking/
[574] http://www.osha.gov/dts/hazardalerts/hydraulic_frac_hazard_alert.html
[575] http://www.jacksonkelly.com/jk/pdf/C2110376.PDF
[576] http://www.epa.gov/airquality/oilandgas/pdfs/20120417fs.pdf
[577] http://www.epa.gov/airquality/oiland gas
[578] *Ibid.*
[579] http://www.npr.org/2012/04/18/150890137/epa-to-slash-air-pollution-from-natural-gas-wells
[580]
http://www.alternet.org/environment/149760/oil_and_gas_companies_illegally_using_diesel_in_fracking_
[581]
http://energycommerce.house.gov/Press_111/20100218/hydraulic_fracturing_memo.pdf
[582] http://ecowatch.org/2012/still-not-regulated/
[583] *Ibid.*
[584] *Ibid.*
[585] http://www.epa.gov/oms/fuels/dieselfuels/index.htm
[586] http://www.dieselforum.org/news/advancements-in-clean-diesel-technology-and-fuel-to-continue-major-reductions-in-black-carbon-emissions
[587] http://www.sfgate.com/bayarea/article/Chevron-fire-truck-didn-t-spark-fire-3794857.php
[588] http://www.sustainableotsego.org/Risk Assessment Natural Gas Extraction-1.htm
[589] http://ecowatch.org/2012/fracking-children/
[590] http://www.dec.ny.gov/data/dmn/rdsgeisfull0911.pdf.
[591] http://marcellusdrilling.com/2012/02/railamerica-buys-wellsboro-corning-short-line-railroad/

[592] http://www.progressiverailroading.com/rail_industry_trends/article/Gas-exploration-and-drilling-in-the-Marcellus-Shale-means-more-traffic-for-Class-Is-short-lines--29103

[593] http://marcellusdrilling.com/2010/03/six-short-line-railroads-in-central-pa-report-business-is-up-40-percent-because-of-marcellus-drilling/

[594] http://www.trefis.com/stock/nsc/articles/137803/how-railroad-companies-could-benefit-from-shale-gas-boom/2012-08-09

[595] http://marcellusdrilling.com/2012/02/csx-other-railroads-get-boost-from-shale-gas-shipments/

[596] *Ibid.*

[597]

http://www.dep.state.pa.us/dep/deputate/airwaste/aq/cars/docs/Final_Act_124_Fact_Sheet.pdf

[598]

http://www.thetrucker.com/News/Stories/2009/2/9/Pennsylvaniatruckidlinglawnowineffect.aspx

[599] http://www.uticaod.com/news/x133071260/Environmental-group-says-pollution-is-streaming-from-fracking-sites

[600]

http://online.wsj.com/article/SB10001424127887323291704578199751783044798.html

[601] http://ec.europa.eu/environment/integration/energy/pdf/fracking study.pdf

[602] http://www.noiseandhealth.org/article.asp?issn=1463-1741;year=2004;volume=6;issue=23;spage=3;epage=20;aulast=Castelo

[603] http://airportnoiselaw.org/dblevels.html

[604]

http://www.alternet.org/water/150794/trailer_talk's_frack_talk%3A_why_a_mayor_was_forced_to_leave_his_town_because_of_gas_drilling_/?page=entire

[605] http://coloradoindependent.com/121266/fracking-operation-in-erie-begins-near-two-elementary-schools-wakes-up-neighborhood

[606] *Ibid.*

[607] http://news.nationalgeographic.com/news/2010/10/101022-energy-marcellus-shale-gas-environment/

[608] http://www.undeerc.org/bakken/bakkenformation.aspx

[609] http://www.energyfromshale.org/bakken-shale-gas

[610] http://www.undeerc.org/bakken/developmenthistory.aspx

[611] http://www.aei-ideas.org/2012/12/americas-economic-miracle-state-north-dakota-sets-new-records-in-october-with-exponential-growth-of-oil-output/

[612] http://www.exxonmobilperspectives.com/2012/09/19/a-hundred-fold-increase-in-oil-production/

[613] http://oilshalegas.com/bakkenshale.html

[614]

http://www.eia.gov/dnav/pet/hist/LeafHandler.ashx?n=pet&s=mcrfpnd1&f=m

[615] http://www.nd.gov/ndic/ic-press/bakken-form-06.pdf

[616] http://oilshalegas.com/bakkenshale.html

[617] http://www.investmentu.com/2011/September/natural-gas-flaring.html

[618] *Ibid.*

[619] http://www.npr.org/2011/09/25/140784004/new-boom-reshapes-oil-world-rocks-north-dakota
[620] http://www.bls.gov/opub/ted/2012/ted_20121126.htm
[621] http://www.aei-ideas.org/2012/12/americas-economic-miracle-state-north-dakota-sets-new-records-in-october-with-exponential-growth-of-oil-output/
[622]
http://ecofriendlydevelopment.net/investment_files/Brian%20Williams%20broadcast,%20N.%20Dakota%27s%20Housing%20Shortage.pdf
[623] http://money.cnn.com/2011/09/28/pf/north_dakota_jobs/index.htm
[624] http://www.npr.org/2011/09/25/140784004/new-boom-reshapes-oil-world-rocks-north-dakota
[625] http://www.npr.org/2011/12/02/142695152/oil-boom-puts-strain-on-north-dakota-towns
[626] http://www.ndwheat.com/buyers/default.asp?ID=295
[627] http://www.minotmilling.com/durum/durum.html
[628] http://www.ndwheat.com/buyers/default.asp?ID=295
[629] http://marcelluseffect.blogspot.com/2012/02/canadian-farmers-call-for-fracking.html
[630]
http://www.marcellusfieldguide.org/index.php/guide/pre_development_issues/effects_on_agriculture/
[631] *Ibid.*
[632] http://pubs.usgs.gov/of/2012/1154/of2012-1154.pdf
[633] http://pubs.cas.psu.edu/FreePubs/PDFs/ee0020.pdf
[634]
http://www.marcellusfieldguide.org/index.php/guide/pre_development_issues/effects_on_agriculture/
[635] http://www.dec.ny.gov/energy/46288.html
[636] http://www.investmentu.com/2011/September/natural-gas-flaring.html
[637] http://www.dec.ny.gov/energy/46288.html
[638] http://pubs.usgs.gov/of/2012/1154/of2012-1154.pdf
[639] http://www.thenation.com/article/171504/fracking-our-food-supply
[640] http://www.sustainableotsego.org/Risk Assessment Natural Gas Extraction-1.htm
[641]
http://63.134.196.109/documents/mpactsofGasDrillingonHumanandAnimalHealth.pdf
[642] http://fractoids.blogspot.com/
[643] http://www.desmogblog.com/%E2%80%98energy-depth%E2%80%99-was-created-major-oil-and-gas-companies-according-industry-memo
[644] http://truth-out.org/news/item/13058-why-are-cows-tails-dropping-off
[645] http://www.thenation.com/article/171504/fracking-our-food-supply
[646] http://www.fwweekly.com/2009/10/14/sacrificed-to-shale/
[647] *Ibid.*
[648]
http://www.alternet.org/water/150794/trailer_talk's_frack_talk%3A_why_a_mayor_was_forced_to_leave_his_town_because_of_gas_drilling_/?page=entire

[649] http://www.theintelligencer.net/page/content.detail/id/561195/Consol-Sued-for-Dunkard-Creek-Fish-Kill.html?nav=515
[650] http://www.tpwd.state.tx.us/landwater/water/environconcerns/hab/ga/
[651] http://www.theintelligencer.net/page/content.detail/id/561195/Consol-Sued-for-Dunkard-Creek-Fish-Kill.html?nav=515
[652] http://www.post-gazette.com/stories/news/us/consol-to-pay-55m-for-clean-water-act-violations-286950/
[653] http://www.eenews.net/public/Landletter/2010/10/21/1
[654] http://thetimes-tribune.com/news/gas-drilling/after-blowout-most-evacuated-families-return-to-their-homes-in-bradford-county-1.1135253
[655] http://www.propublica.org/article/science-lags-as-health-problems-emerge-near-gas-fields/single
[656] http://www.water-contamination-from-shale.com/louisiana/louisiana-fracking-operations-suspect-in-cow-deaths/
[657] http://www.facebook.com/#!/Realpromisedland/info
[658] http://www.usatoday.com/news/nation/2010-12-14-1Alouisiana14_CV_N.htm
[659] http://lancasterfarming.com/news/northeedition/Couple-Reeling-From-Natural-Gas-Mess-
[660] *Ibid.*
[661] *Ibid.*
[662] http://www.ahs.dep.pa.gov/eFACTSWeb/searchResults_singleViol.aspx?InspectionID=2115722
[663] http://www.ahs.dep.pa.gov/eFACTSWeb/searchResults_singleViol.aspx?InspectionID=2076777
[664] http://www.ahs.dep.pa.gov/eFACTSWeb/searchResults_singleViol.aspx?InspectionID=2095088.
[665] http://www.observer-reporter.com/article/20121218/NEWS04/121219324
[666] http://www.propublica.org/article/injection-wells-the-poison-beneath-us
[667] http://www.marcellusgas.org/graphs/PA#prodollars
[668] http://www.epa.gov/radon/pubs/citguide.html
[669] http://www.desmogblog.com/radionuclides-tied-shale-gas-fracking-can-t-be-ignored-possible-health-hazard
[670] http://www.marcellus-shale.us/radioactive-shale.htm
[671] http://www.ohio.com/blogs/drilling/ohio-utica-shale-1.291290/study-says-pennsylvania-drilling-waste-high-in-radium-1.331703
[672] http://www.frackcheckwv.net/
[673] http://www.dispatch.com/content/stories/local/2012/09/03/gas-well-waste-full-of-radium.html
[674] http://www.nytimes.com/2011/02/27/us/27gas.html?pagewanted=all
[675] http://www.energyforamerica.org/2012/11/29/hollywood-perpetuates-hydraulic-fracturing-myths-in-promised-land/
[676] http://www.nrdc.org/energy/files/Fracking-Wastewater-FullReport.pdf

677 http://www.pennfuture.org/userfiles12/EnvironmentalHearingBoardNOA201
2-10-01.pdf

678 http://www.post-gazette.com/stories/local/region/pennfuture-accuses-dep-
of-permit-dishonesty-656236/

679 http://www.post-gazette.com/stories/local/region/pennfuture-accuses-dep-
of-permit-dishonesty-656236/

680 http://www.post-gazette.com/stories/local/region/pennfuture-accuses-dep-
of-permit-dishonesty-656236/

681 *Ibid.*

682 http://pennbpc.org/sites/pennbpc.org/files/CMSC-Final-Report.pdf

683 http://thetimes-tribune.com/news/gas-company-whistle-blower-details-
spills-errors-1.1234817

684 http://insideclimatenews.org/news/20120515/bureau-land-management-
blm-fracking-regulations-natural-gas-chemical-disclosure?page=show

685 http://newamericamedia.org/2012/05/feds-punt-on-leadership-over-
fracking-rules-experts-say.php

686 http://www.nrdc.org/media/2012/120509.asp

687 *Ibid.*

688 http://www.waterworld.com/articles/2011/12/wastewater-plants-not-
designed-for-fracking-water-says-robert-f-kennedy-jr.html

689
http://s3.amazonaws.com/propublica/assets/monongahela/MarcellusShaleWat
erManagementChallenges 11.08.pdf

690 http://old.post-gazette.com/pg/08322/928571-113.stm

691 http://www.propublica.org/article/wastewater-from-gas-drilling-boom-may-
threaten-monongahela-river

692
http://www.portal.state.pa.us/portal/server.pt/community/newsroom/14287?id
=%2017071%20&typeid=1

693 http://www.epa.gov/region3/marcellus_shale/pdf/letter/krancer-letter5-12-
11.pdf

694 http://water.epa.gov/type/groundwater/uic/class2/

695 http://www.propublica.org/article/injection-wells-the-poison-beneath-us

696 *Ibid.*

697 *Ibid.*

698 http://stateimpact.npr.org/pennsylvania/maps/location-of-deep-injection-
wells-in-pennsylvania/

699 http://www.lcountyfracking.org/archives/465

700 http://www.propublica.org/article/an-unseen-leak-then-boom

701 http://www.hutchnews.com/print/Sun--explosions-reflection--1

702 http://www.rrc.state.tx.us/about/faqs/hydraulicfracturing.php

703 http://www.thestar.com/business/article/1159854

704 http://www.ibtimes.com/articles/342886/20120518/earthquake-texas-
hydraulic-fracturing-waste-water-injection.htm

705
http://i2.cdn.turner.com/cnn/2012/images/06/15/induced.seismicity.prepublica
tion.pdf

[706] *Ibid.*
[707] http://www.nature.com/news/method-predicts-size-of-fracking-earthquakes-1.9608
[708] http://www.cnn.com/2012/06/15/us/fracking-earthquakes/index.html
[709] http://www.desmogblog.com/directory/vocabulary/6566
[710] Rubinstein, Justin L., *et al.*, "Present Triggered Seismicity Sequence in the Raton Basin of Southern Colorado/Northern New Mexico," professional paper presented to the American Geophysical Union Fall meeting, Dec. 3-7, 2012.
[711] http://www.cuadrillaresources.com/cms/wp-content/uploads/2011/11/Final_Report_Bowland_Seismicity_02-11-11.pdf
[712] http://www.bcogc.ca/document.aspx?documentID=1270&type=.pdf
[713] http://www.texassharon.com/2012/07/13/three-earthquakes-last-night/
[714] http://redgreenandblue.org/2011/11/06/did-fracking-cause-the-oklahoma-earthquake/
[715] http://www.ibtimes.com/articles/342886/20120518/earthquake-texas-hydraulic-fracturing-waste-water-injection.htm
[716] *Ibid.*
[717] http://www.scientificamerican.com/article.cfm?id=did-fracking-cause-oklahomas-largest-recorded-earthquake
[718] http://www.okgeosurvey1.gov/pages/earthquakes/information.php
[719] http://www.occeweb.com/STRONGER REVIEW-OK-201-19-2011.pdf
[720] http://www.ogs.ou.edu/pubsscanned/openfile/OF1_2011.pdf
[721] http://www.allgov.com/Controversies/ViewNews/Arkansas_Suspends_Drilling_of_Injection_Wells_after_Earthquake_Swarm_110302
[722] http://www.ibtimes.com/articles/342886/20120518/earthquake-texas-hydraulic-fracturing-waste-water-injection.htm
[723] http://www.aogc.state.ar.us/Hearing Orders/2011/July/180A-2-2011-07.pdf
[724] http://ohiodnr.com/downloads/northstar/UICReport.pdf
[725] http://truth-out.org/news/item/10606-special-investigation-the-earthquakes-and-toxic-waste-of-ohios-fracking-boom
[726] http://www.cnn.com/2012/06/15/us/fracking-earthquakes/index.html
[727] http://truth-out.org/news/item/10606-special-investigation-the-earthquakes-and-toxic-waste-of-ohios-fracking-boom
[728] http://truth-out.org/index.php?option=com_k2&view=item&id=7245:regulators-say-fracking-wastewater-well-caused-12-earthquakes-in-ohio
[729] http://www.utexas.edu/news/2012/08/06/correlation-injection-wells-small-earthquakes/
[730] *Ibid.*
[731] http://www.nrdc.org/media/2012/120509.asp
[732] *Ibid.*
[733] http://www.edf.org/energy/getting-natural-gas-right
[734] http://m.npr.org/story/140872251
[735] http://63.134.196.109/documents/mpactsofGasDrillingonHumanandAnimalHealth.pdf

736 http://www.post-gazette.com/stories/local/marcellusshale/washington-county-families-sue-over-fracking-water-testing-637631/?print=1
737 *Ibid.*
738 *Ibid.*
739 http://canon-mcmillan.patch.com/blog_posts/the-pennsylvania-dep-another-red-herring-2
740

http://epw.senate.gov/public/index.cfm?FuseAction=Minority.PressReleases&ContentRecord_id=fd161ae5-802a-23ad-410b-549dd302203b
741 http://fuelfix.com/blog/2011/12/08/epa-says-hydraulic-fracturing-polluted-groundwater/
742

http://www.opensecrets.org/politicians/industries.php?type=C&cid=N0000558
2&newMem=N&cycle=2012
743

http://www.opensecrets.org/politicians/industries.php?cycle=Career&type=C&cid=N00005582&newMem=N&recs=20
744 http://trib.com/opinion/editorial/epa-s-silence-does-a-disservice-to-wyoming/article_0921b4ec-3d86-5a6e-bd5b-67bb739c138a.html
745 http://trib.com/opinion/columns/finding-answers-for-pavillion-residents/article_2d4e5b10-4399-5e45-9f5e-ee75d8949de8.html
746

http://science.house.gov/sites/republicans.science.house.gov/files/documents/hearings/HHRG-112-SY20-20120201-SD001.pdf
747 http://www.gastruth.org/?cat=7
748

http://files.dep.state.pa.us/AboutDEP/AboutDEPPortalFiles/RemarksAndTestimonies/MLK-Testimony-111611.pdf
749 http://www.nicholas.duke.edu/cgc/pnas2011.pdf
750 http://articles.philly.com/2011-12-02/news/30467569_1_drinking-water-water-resources-methane
751 http://www.post-gazette.com/stories/local/marcellusshale/chamber-officials-lobby-to-keep-shale-costs-low-645818/
752

http://www.sec.gov/Archives/edgar/data/858470/000104746912001751/a2207418z10-k.htm
753 http://www.americanprogress.org/issues/2012/01/exxonmobil_profits.html
754

http://topics.nytimes.com/top/news/business/companies/exxon_mobil_corporation/index.html
755 http://www.cfr.org/united-states/new-north-american-energy-paradigm-reshaping-future/p28630
756 http://www.dummies.com/how-to/content/true-conspiracy-the-ford-pinto-memorandum.html
757 http://www.motherjones.com/politics/1977/09/pinto-madness?page=1
758

http://www.earthisland.org/journal/index.php/elist/eListRead/robert_kennedy_jr._on_fracking_media_and_changing_light_bulbs

759 http://www.nytimes.com/2012/10/01/nyregion/with-new-delays-a-growing-sense-that-gov-andrew-cuomo-will-not-approve-gas-drilling.html?pagewanted=all&_r=0

760 http://newyork.newsday.com/news/region-state/cuomo-fracking-regulations-likely-delayed-into-2013-1.4246369

761 http://www.energyindepth.org/tag/pa/

762 http://eidmarcellus.org/marcellus-shale/artists-against-fracking-no-artists-looking-for-relevance-part-2/12534/

763 http://eidmarcellus.org/marcellus-shale/artists-against-fracking-no-artists-looking-for-relevance-part-i/12509/

764 *Ibid.*

765 http://www.cnn.com/2012/08/29/us/new-york-fracking-artists-protest/index.html

766 http://artistsagainstfracking.com/our-trip/

767 http://bigstory.ap.org/article/yoko-ono-sean-lennon-tour-pa-gas-drilling-sites

768 http://susquehannacounty.wnep.com/news/news/141213-celebrity-fracking-bus-tour-causes-commotion

769 http://bigstory.ap.org/article/yoko-ono-sean-lennon-tour-pa-gas-drilling-sites

770 http://susquehannacounty.wnep.com/news/news/141213-celebrity-fracking-bus-tour-causes-commotion

771 http://www.variety.com/review/VE1117941971/

772 http://www.cfr.org/united-states/new-north-american-energy-paradigm-reshaping-future/p28630

773 http://www.energyindepth.org/wp-content/uploads/2011/11/Debunking-Gasland.pdf

774 http://fractoids.blogspot.com/

775 http://lancasteronline.com/article/local/361158_State-geologist-discusses-Marcellus-Shale-in-talk-here.html

776 http://lancasteronline.com/article/local/362603_Pa--official-apologizes-for-Nazi-propaganda-comment-made-here.html

777 http://johnhanger.blogspot.com/2011/02/gasland-and-oscars.html

778 http://www.truthlandmovie.com/

779 http://blog.littlesis.org/2012/06/13/fracking-industrys-answer-to-gasland-devised-by-astroturf-lobbying-group-and-political-ad-agency/

780 http://commonsense2.com/2012/09/naturalgasdrilling/connecting-the-dots-part-10-let%e2%80%99s-go-to-the-movies/

781 http://blog.littlesis.org/2012/06/13/fracking-industrys-answer-to-gasland-devised-by-astroturf-lobbying-group-and-political-ad-agency/

782 http://fracktoids.blogspot.com/2012/07/energy-in-depth-dots.html?showComment=1341985781003

783 http://commonsense2.com/2012/05/unsung-heroes/meet-the-1st-lady-of-fractivism-the-woman-who-connects-the-dots/

784 http://commonsense2.com/2012/09/naturalgasdrilling/connecting-the-dots-part-10-let%e2%80%99s-go-to-the-movies/

785 http://www.prnewswire.com/news-releases/fracknation-premieres-january-22-on-axs-tv-183812371.html

[786] http://www.imdb.com/title/tt1558250/
[787] http://www.boxofficemojo.com/movies/?id=promisedland2012.htm
[788] http://www.boxofficemojo.com/movies/?id=promisedland2012.htm
[789] http://www.movieinsider.com/m9828/promised-land/production/
[790] http://www.huffingtonpost.com/robert-f-kennedy-jr/fracking-movie-promised-land_b_2251339.html
[791] http://www.foxnews.com/opinion/2012/10/01/hollywood-hypocrisy-and-matt-damon-anti-fracking-film/
[792] http://www.nationalreview.com/planet-gore/328561/ithe-bourne-stupidityi-matt-damon-takes-fracking-greg-pollowitz
[793] http://www.breitbart.com/Big-Hollywood/2012/09/27/Matt-Damon-s-Anti-Fracking-Movie-Goes-Full-Conspiracy-Theory
[794] http://www.cnn.com/2013/01/02/opinion/bennett-fracking-movie/index.html
[795] http://www.facebook.com/#!/Realpromisedland/info
[796] http://blog.heritage.org/2012/09/28/matt-damons-anti-fracking-movie-financed-by-oil-rich-arab-nation/
[797] http://johnhanger.blogspot.com/2013/01/promised-land-fracks-or-blows-it-my.html
[798] http://www.participantmedia.com/wp-content/uploads/2011/05/Participant-Media-Our-History.pdf
[799] http://www.businessinsider.com/matt-damon-movie-blasts-fracking-backed-by-uae-2012-9#ixzz2Ha60FQfw
[800] http://cnsnews.com/node/616420
[801] http://marcelluscoalition.org/2013/01/msc-engages-public-with-natural-gas-facts-counters-promised-land/
[802] http://energy.aol.com/2012/09/12/marcellus-gas-output-jumps-despite-low-price/
[803] http://marcellusdrilling.com/2012/10/detailed-look-at-pas-first-half-production-numbers/
[804] http://www.eia.gov/dnav/ng/hist/rngwhhdM.htm
[805] http://www.nytimes.com/2012/10/21/business/energy-environment/in-a-natural-gas-glut-big-winners-and-losers.html?pagewanted=all&_r=0
[806] *Ibid.*
[807] http://www.nytimes.com/2012/10/21/business/energy-environment/in-a-natural-gas-glut-big-winners-and-losers.html?pagewanted=all&_r=0
[808] http://science.howstuffworks.com/environmental/green-science/clean-coal.htm
[809] http://energy-facts.org/?utm_source
[810] *Ibid.*
[811] http://encyclopedia2.thefreedictionary.com/United+Mine+Workers
[812] http://sierraclub.typepad.com/compass/2012/02/the-sierra-club-and-natural-gas.html
[813] http://www.sierraclub.org/sierra/201207/pennsylvania-fracking-shale-gas-200-sidebar.aspx
[814] http://ec.europa.eu/dgs/jrc/index.cfm
[815] http://ec.europa.eu/dgs/jrc/index.cfm
[816] http://rt.com/news/solar-energy-record-break-332/
[817]

http://insideclimatenews.org/news/20121220/clean-energy-economy-2012-year-end-solyndra-wind-energy-solar-power-feed-in-tariff-germany-low-carbon-california?page=show

818 *Ibid.*

819 http://money.msn.com/business-news/article.aspx?feed=SALP&date=20120807&id=15423276

820 http://www.examiner.com/article/drilling-permits-decline-sharply-for-the-pennsylvania-marcellus-formation

821 http://articles.philly.com/2012-07-08/business/32589447_1_natural-gas-prices-drilling-natural-gas/2

822

http://www.thereporteronline.com/article/20121001/NEWS03/121009986/next-cold-war-gas-drilling-boom-rattles-russia&pager=full_story

823 http://knappap.blogspot.com/2012/01/wet-gas-vs-dry-gas-do-you-know-what.html

824 *Ibid.*

825

http://www.canada.com/business/Asian+market+looking+natural+executive+director+says/7095235/story.html

826

http://www.canada.com/business/Asian+market+looking+natural+executive+director+says/7095235/story.html

827 http://www.jdsupra.com/legalnews/us-chamber-of-commerce-sues-sec-to-ove-75416/

828 http://www.prwatch.org/news/2012/10/11802/big-oil-and-us-chamber-fight-keep-foreign-bribery-flourishing

829 http://www.psehealthyenergy.org/data/LNG_SignOnLetterPDF.pdf

830 http://www.psehealthyenergy.org/events/view/144

831 http://www.magazine.columbia.edu/print/1091

832 http://www.examiner.com/article/drilling-permits-decline-sharply-for-the-pennsylvania-marcellus-formation

833 http://www.magazine.columbia.edu/print/1091

834 http://www.villagevoice.com/2012-09-19/news/boom-or-doom-fracking-environment

PHOTO: Diane Slegmund

Rail cars carrying sand and chemicals to a natural gas site pass close to a child care center near Wyalusing, Pa.

INDEX

ABOUT THE AUTHOR

Walter M. Brasch, Ph.D., is an award-winning social issues journalist and the author of 16 other books, most of which fuse historical and contemporary social issues. Among his books is *Before the First Snow*, a critically-acclaimed novel that looks at what happens when government and energy companies form a symbiotic relationship, using 'cheaper, cleaner' fuel and the lure of jobs in a depressed economy but at the expense of significant health and environmental impact.

He is professor emeritus of mass communications and journalism, and a former newspaper and magazine reporter and editor, and multi-screen multi-media writer–producer.

He is vice-president of the Central Susquehanna chapter of the ACLU, vice-president and co-founder of the Northeast Pennsylvania Homeless Alliance, a member of the board of the Keystone Beacon Community for healthcare coordination, and is active in numerous social causes. He was a Commonwealth Speaker for the Pennsylvania Humanities Council, and was active in emergency management.

Dr. Brasch is featured columnist for *Liberal Opinion Week*, senior correspondent for the *American Reporter*, senior editor for *OpEdNews*, and an editorial board member of the *Journal of Media Law and Ethics*.

He was president of the Pennsylvania Press Club and the Keystone State professional chapter of the Society of Professional Journalists, vice-president of the Pennsylvania Women's Press Association, and founding coordinator of Pennsylvania Journalism Educators. He is a member of the National Society of Newspaper Columnists, the Authors Guild, and The Newspaper Guild (CWA/AFL-CIO). He is listed in *Who's Who in America, Contemporary Authors, Who's Who in the Media*, and *Who's Who in Education*.

He was recognized in 2012 by the Pennsylvania Press Club with the Communicator of Achievement award for lifetime achievement in journalism and public service. Among recent writing awards are multiple awards from the National Society of Newspaper Columnists, Society of Professional Journalists, National Federation of Press Women, USA Book News, Independent Book Publishing Professionals Group, Pennsylvania Press Club, Pennsylvania Women's Press Association, Penn-Writers, Pacific Coast Press Club, Press Club of Southern California, and the International Association of Business Communicators.

He was honored by San Diego State University as a Points of Excellence winner in 1997. At Bloomsburg University, he earned the Creative Arts Award, the Creative Teaching Award, and was named an Outstanding Student Advisor. He received the first annual Dean's Salute to Excellence in 2002, and a second award in 2007, and the Maroon and Gold Quill Award for nonfiction. He is the 2004 recipient of the Martin Luther King Jr. Humanitarian Service Award.

Dr. Brasch earned an A.B. in sociology/social welfare from San Diego State College, an M.A. in journalism from Ball State University, and a Ph.D. in mass communication/journalism, with cognate areas in both American government/public policy and language and culture studies, from The Ohio University.

He is married, has two children, and is surrounded by animals and the rural beauty of northeastern Pennsylvania.

To learn more about Dr. Brasch, visit
http://www.walterbrasch.com

PHOTO: Margaret Lettis

The rigs will come down. The wells will be capped. But the effects from fracking will continue for decades.

CPSIA information can be obtained at www.ICGtesting.com
Printed in the USA
LVOW101618140513

333776LV00016B/701/P